I0065334

IET MATERIALS, CIRCUITS AND DEVICES SERIES 69

Fibre Bragg Gratings in Harsh and Space Environments

Other volumes in this series:

Fibre Bragg Gratings in Harsh and Space Environments

Principles and applications

Brahim Aïssa, Emile I. Haddad, Roman V. Kruzelecky, and Wes R. Jamroz

The Institution of Engineering and Technology

Published by The Institution of Engineering and Technology, London, United Kingdom

The Institution of Engineering and Technology is registered as a Charity in England & Wales (no. 211014) and Scotland (no. SC038698).

© The Institution of Engineering and Technology 2019

First published 2019

This publication is copyright under the Berne Convention and the Universal Copyright Convention. All rights reserved. Apart from any fair dealing for the purposes of research or private study, or criticism or review, as permitted under the Copyright, Designs and Patents Act 1988, this publication may be reproduced, stored or transmitted, in any form or by any means, only with the prior permission in writing of the publishers, or in the case of reprographic reproduction in accordance with the terms of licences issued by the Copyright Licensing Agency. Enquiries concerning reproduction outside those terms should be sent to the publisher at the undermentioned address:

The Institution of Engineering and Technology
Michael Faraday House
Six Hills Way, Stevenage
Herts, SG1 2AY, United Kingdom

www.theiet.org

While the authors and publisher believe that the information and guidance given in this work are correct, all parties must rely upon their own skill and judgement when making use of them. Neither the authors nor publisher assumes any liability to anyone for any loss or damage caused by any error or omission in the work, whether such an error or omission is the result of negligence or any other cause. Any and all such liability is disclaimed.

The moral rights of the authors to be identified as authors of this work have been asserted by them in accordance with the Copyright, Designs and Patents Act 1988.

British Library Cataloguing in Publication Data
A catalogue record for this product is available from the British Library

ISBN 978-1-78561-980-9 (hardback)
ISBN 978-1-78561-981-6 (PDF)

Typeset in India by MPS Limited
Printed in the UK by CPI Group (UK) Ltd, Croydon

Contents

List of figures

List of tables

Preface

Fibre optic sensor (FOS) technology has been under increasing development since more than 60 years. It has resulted in the production of various advanced devices, including sensors of temperature, pressure, vibration; fibre optic gyroscopes and chemical probes. FOSs offer various advantages, for example, their improved sensitivity as compared to existing techniques and geometric versatility, which permits their configuration with different shapes. In addition, because the materials devices are dielectric, FOSs can be used in harsh environment conditions, e.g., high voltage, high temperature or corrosive media. Besides, these sensors are compatible with communications systems and have the capacity to carry out remote sensing.

FOS has been used to enhance and test the integrity, efficiency, safety, and durability of structures, vehicles, medical devices, and more across a multitude of industries. Recent advancements have enabled FOS to expand its abilities to include new fields across applications in medical, energy and space. This is helping engineers solve problems they are faced with today and innovate their designs.

There are a vast number of real-world implications for fibre optic technology, as well as a realm of possibilities for the future. Basically, fibre Bragg gratings represent an important element in the emerging fields of optical communications and optical sensing. Despite its vast usefulness, the device is comparatively simple. In its simplest form, a fibre Bragg grating consists of a periodic modulation of the refractive index in a core of a single mode optical fibre, where the phase fronts are perpendicular to the fibre's longitudinal axis and with grating planes having a constant period. Light, guided along the core of an optical fibre, is scattered by each grating plane. If the Bragg condition is satisfied, the contributions of reflected light from each grating plane add constructively in the backward direction to form a back-reflected peak with centre wavelength defined by the grating period.

This book is dedicated to address the critical challenge of developing FBG for applications that require operation in harsh environments and focuses on space-frontiers application. It covered all the aspects of the technology, i.e. from basic research through design, fabrication and testing to the industrial implementation of high temperature and radiation-resistant optical fibres. The areas of testing encompassed high-temperature annealing and effect of the exposure to gamma radiation encountered by satellites.

The book starts with providing a detailed introduction to the key element of the FOS, namely, fibre Bragg gratings (FBGs) that include the fundamental under-standing of the spectral properties of the fibres used. Then, some of the most recent developments in the application of FBGs sensors for extreme environment conditions are reviewed. Although the dominant technologies used today to perform temperature measurements still are based on electrical sensors, optical sensors were shown to offer a promising alternative to challenging applications.

Harsh environments, distributed systems, space and long-term deployments are typical examples where the characteristics of the FBGs systems can provide the clear advantage and a highly effective solution as compared to conventional electrical sensors.

The introduction in **Chapter 1** is followed by a review of the technology of Bragg gratings in optical fibres, where the phenomenon of photosensitivity in optical fibres was introduced and detailed in **Chapter 2**. While the **Harsh environment FBGs sensing** is discussed in **Chapter 3**, the properties of Bragg gratings are examined, and some of the important developments in devices and applications are presented in **Chapter 4**. The most common fabrication techniques (interferometric, phase mask and point by point) are described in **Chapter 5** with reference to the advantages and the disadvantages in utilising them for inscribing Bragg gratings.

The last chapters of the book are oriented towards space applications. **Chapter 6** highlights deals with the presence in the space of micrometeoroids and orbital debris that are characterizing the lower earth orbit (LEO), which in turn presents a true hazard to orbiting satellites. We addressed a short review of the space debris challenges and reported on the evaluation of FBG sensors in the LEO environment. Then the most recent experimental results are presented on the applications of FBG with self-healing composite materials used in space. In **Chapter 7**, we have presented the fibre sensor demonstrator (FSD) flight validations on ESA's PROBA-s spacecraft, as launched in November 2009 and still in operation until now. This technology has proven indeed to provide reliable long-term operations in harsh environments with significant benefits relative to relevant electronic sensors and signal processing in terms of immunity to EMI/EMC radiation, low-loss and lightweight fibre-optic signal harness, sensor mass and signal quality.

Finally, in **Chapter 8**, we reviewed the integration of fibre optics with FBGs with MEMS actuators to facilitate various tunable fibre-optic and integrated-optic devices for optical signal processing, such as variable time delay lines and the sensing of physical parameters, such as temperature, pressure and acceleration/tilt, while **Chapter 9** concludes with a summary and challenges of the FOS technology.

The book is amended with an extensive and updated survey of the published articles and conference reports. It is hoped that the book will be of interest to those involved in the investigation of FOS technology at large, ranging from novice students to the most experienced end users. The intention is also to stimulate debate and reinforce the importance of a multidisciplinary approach in this exciting field.

Acknowledgements

First and foremost, we would like to express our gratitude to the many people who saw us through this book; to all those who provided support, talked things over, read, wrote, offered comments, allowed us to quote their remarks and assisted in the editing, proofreading and design, to all who have either directly or indirectly contributed to this book.

We would like to offer our special thanks to Ms Jane Bachynski, President of MPB Communications Inc., who has provided a unique positive environment where the business objectives are realized through the cultivation of innovation and by harnessing scientific curiosity.

We would like to express our gratitude to our colleagues from various institutions and organizations who provided us with assistance to our work.

Especially we thank Darius Nikanpour, Stéphane Gendron and Philip Melanson from the Canadian Space Agency; Daniel Therriault from École Polytechnique de Montreal; Mourad Nedil from University of Quebec in Abitibi Temiscamingue; Mohamad Asgar and S.V. Hoa from Concordia University; Philippe G. Merle from the Canadian Department of National Defense; Jason Loiseau, Jimmy Verreault and Andrew Higgins from McGill university; Christopher Semprimoschnig, Iain McKenzie, Nikos Karafolas, Pierrik Vuilleumier and Philippe Poinas from the European Space Agency; Francesco Ricci, Joshua Lamorie and Eric Edwards from Xiphos Technologies Inc.

We wish to acknowledge the help provided by Kamel Tagziria, Jing Zou, Najeeb Mohammed, Jonathan Lavoie, Alireza Nakhaei and Roy Josephs from MPB Communications Inc. for their assistance with experimental data.

We wish to underline the valuable help provided by the staff of The Institution of Engineering and Technology (IET) publishing group, with a special mention to Sarah Lynch, Olivia Wilkins and Paul Deards for their availability and willingness to provide guidance and advice during the preparation of the manuscript.

We would like to gratefully acknowledge the financial and technical contribution of the Canadian Space Agency, the Natural Science and Engineering Research Council (NSERC) of Canada and the European Space Agency. Dr. Brahim Aïssa would like to express a special thank to Qatar Environment and Energy Research Institute (Hamad Bin Khalifa University, Qatar Foundation).

Abbreviations

AC	Alternating current (AC)
AD	Atomic displacement
AO	Atomic oxygen
ASE	Amplified spontaneous emission
ASTM	American Society for Testing and Materials
BBO	Beta-barium borate
BOTDA	Brillouin optical time domain analysis
CCD	Charged coupled device
CCGs	Chemical-composition gratings
CDMA	Code-division multiple access
CFRP	Carbon fibre reinforced polymers
CIMU	Current inertial measurement units
CMOS	Complementary metal oxide semiconductor
CNT	Carbon nanotube
COD	1,5-Cyclooctadiene
CSA	Canadian Space Agency
CW	Continuous wave
CWL	Centre wavelength
DAC	Digital to analogue
DC	Direct current
DCPD	Dicyclopentadiene
DCU	Dispersion compensation unit
DFB	Distributed feedback
DMCP	Di(methylcyclopentadiene)
DNA	Deoxyribonucleic acid
DRIE	Deep reactive ion etching to facilitate
DTG	Draw tower grating
DUT	Device under Test
ECSS	European Cooperation for Space Standardization
EDFA	Erbium-doped fibre amplifier
EDS	Energy dispersive
EM	Electromagnetic
EMC	Electromagnetic compatibility
EMI	Electromagnetic interference
EN	European Standard
ESA	European Space Agency

ESD	Electrostatic discharge
FBG	Fibre Bragg gratings
FC/APC	Fibre optic connector/angled physical contact
FEA	Finite element analysis
FIR	Fluorescence intensity ratio
FL	Fluorescence lifetime
FORC	Fibre Optic Research Center
FOS	Fibre optic sensor
FP	Fabry-Pérot
FPGA	Field programmable gate array
FRP	Fibre reinforced polymer
FSD	Fibre-optic sensor demonstrator
FSR	Free spectral range
FTTH	Fibre-to-the-home
GIOVE	Galileo In-Orbit Verification
GNSS	Global navigation satellite system
GO	Geostationary orbit
HAP	High-altitude platforms
HB	Highly birefringent
HC	Honeycomb
HRTEM	High-resolution transmission electron microscopy
I/O	Input/Output
I2C	Inter-integrated circuit
IFOG	Integrated fibre-optic gyroscopes
IOFDR	Incoherent optical frequency domain reflectometry
IR	Infrared
ISO	International Standard
ISS	International Space Station
LD	Laser diode
LED	Light emitting diode
LEO	Low earth orbit
LIDAR	LIght detection and ranging
LISA	Laser Interferometer Space Antenna
LNG	Liquid natural gas
LPFG	Long-period fibre gratings
LPG	Long-period grating
LTP	LISA technology pathfinder
LUT	Lookup table
MAMSK	Multi-amplitude minimum shift keying
MCVD	Modified chemical vapour deposition
MEMS	Micro-electro-mechanical-systems
MEO	Medium Earth orbit
MLIB	Multilayer insulation blankets
MMOD	Micrometeorites and orbital debris

MOF	Microstructured optical fibre
Mproof	Mass proof
NA	Numerical aperture
NASA	National Aeronautics and Space Administration
NIEL	Non-ionizing energy loss
NIR	Near infrared
OD	Outer diameter
OFBG	Optical fibre Bragg grating
OFCS	Optical fibre current sensor
OFDM	Orthogonal frequency division multiplexing
OFDR	Optical frequency domain reflectometry
OFS	Optical fibre sensor
OSA	Optical spectral analyser
OTDR	Optical time domain reflectometry
P/T	Pressure/Temperature
PBG	Photonic band gap
PCB	Printed circuit boards
PD	Photodetetor
PDMS	Polydimethylsiloxane(s)
PIC	Photonic integrated circuits
PMMA	Poly(methyl methacrylate)
PMUF	Poly(melamine-urea-formaldehyde)
POTDR	Polarization-optical time domain refractometry
PRARE	Precision range and range-rate equipment
PROBA 2	Project for on-board autonomy satellite
RBM	Radial breathing mode
RF	Radio-frequency
RFI	Radio frequency interference
RGC	ruthenium Grubbs catalyst
RI	Refractive index
RIA	Radiation-induced attenuation
RMSE	Root mean square error
RoF	Radio-over-fibre
ROMP	Ring opening metathesis polymerisation
RPM	Revolution per minute
RTDs	Resistance temperature detectors
SCA	Special centroid algorithm
SCWR	Supercritical water-cooled reactors
SEM	Scanning electron microscopy
SHM	Structural health monitoring
SI	International System of Units
SL	Scanning laser
SLD	Super luminescent diodes
SMF	Single mode fibre

SNR	Signal-to-noise ratio
SOI	Silicon on insulator
SP	Shear panels
SPENVIS	Space Environment Information System
SRAM	Static random access memory
SS	Stainless steel
SSTL	Surrey Satellite Technology Ltd
SWCNT	Single-walled carbon nanotube
TDM	Time-division multiplexing
TEM	Tunnelling electron microscopy
TID	Total ionizing dose
TIR	Total internal reflection
TML	Total mass loss
TRL	Technology readiness level
UV	Ultraviolet
VHTR	Very-high-temperature reactors
VOA	Variable optical attenuator
WDM	Wavelength-division multiplexing
XPS	X-Ray photoelectron spectroscopy
XRD	X-Ray diffractometer
μNAV	Micronavigator
5E2N	5-Ethylidene-2-norbornene
5V2N	5-Vinyl-2-norbornene

Chapter 1

Fundamentals

This chapter is an introduction to the fibre Bragg gratings (FBG) aimed at providing a fundamental understanding of the spectral properties of the optical fibres that are used for the manufacturing of the fibre optic sensors (FOS). The concepts presented in this chapter are important for the analysis and optimisation of their use in harsh environment conditions.

Various types of FOS are employed with different principles that include reflectometry and interferometry, polarisation effect, variation of the refractive index (RI), light intensity modulation and so on. However, the FBG appears of particular interests for a large set of applications. As shown in Figure 1.1, the FBG could be seen as the regular variation of the RI along the longitudinal direction of the core in the fibre optic at selected and well-defined region. FBG is hence reflecting the incident light in a thin and narrow bandwidth which is localised on the Bragg wavelength, namely 'λ_B' that is given as [1]

$$\lambda_B = 2N_{eff}\Lambda \tag{1.1}$$

where 'Λ' is the lateral and/or spatial period of the periodic variation and 'N_{eff}' is the effective index of light that is propagating along a single mode fibre. The spatial period of this variation needs to be determined by the value of the light wavelength aimed to be reflected.

As detailed in Chapter 5, the Bragg grating period is fabricated by intense irradiation of the optical fibre core through ultraviolet (UV) laser light. The targeted pattern can be spatially varied using appropriate mask. Habitually, UV photons below 300 nm are used as they contain high energy able to break easily the silicon oxygen stable bonds of the core, thereby modifying appropriately both its structure and its RI. This intensity variation of the UV laser through the interference of coherent light beams and a mask placed over the irradiated fibre triggers the periodic variation of the RI of the fibre.

The treated optical way (i.e. FBG) of the modified fibre is now used as a light-selective mirror: the specific wavelength which is travelling through the fibre optical tunnel is reflected partially every time the optical index varies, and these reflections are destructive for the most light or wavelengths. The light continues to propagate as long as the fibre is uninterrupted. However, at a specific value of wavelengths, constructive interference phenomenon occurs and the light is reflected back along the optical fibre.

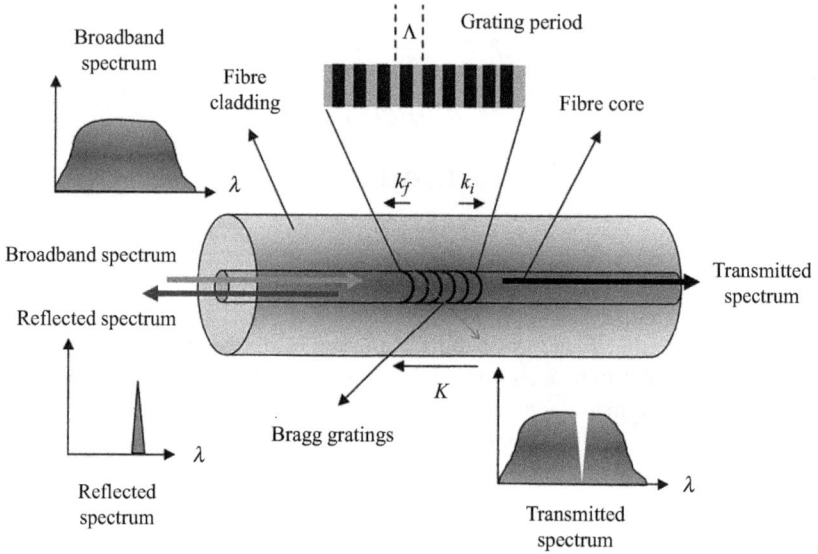

Figure 1.1   Schematic illustration of the principle of FBG (K: grating wave vector, k_i: is wave vector of the incident wave and k_f: wave vector of the scattered radiation)

The FBG somehow summarises all the existing advantages of an optical fibre. This includes their lightweight, sensitivity, multiplexing capability, electrically passive operation and their insensitivity to electromagnetic interference. This makes them in a superior class over conventional optical fibres [2,3] and allows them to be suitable for a large range of advanced applications, namely sensing, filtering, routing and switching [4].

Habitually, an excimer UV laser source is employed to create the FBG in the optical fibre either by means of an internal writing technology [5,6] or through the external one [1]. Despite this, these processing are limited only to the fibre core material that is photosensitive, which is not suitable for high power operation. Only not long ago, chemical modifications were made in a non-photosensitive fibre core material but at the cost of longer UV exposure time [7].

Equation (1.1) shows that the variation of the physical and/or mechanical characteristics of the grating region will affect the reflected wavelength 'λ_B'.

As a matter of fact, mechanical strain on the fibre affects Λ and N_{eff} through the stress-optical effect. In the same way, any change in terms of temperature will involve changes in N_{eff} through the thermo-optical effect and Λ is thereby influenced by the thermal expansion (and/or contraction). This scenario is clearly indicated in (1.2), where the first term expresses the effect of mechanical strain on λ_B and the second term illustrates the effect of temperature:

$$\Delta\lambda_B = \lambda_B(1 - \rho_\alpha)\Delta\varepsilon + \lambda_B(\alpha + \xi)\Delta T \tag{1.2}$$

where '$\Delta\lambda_B$' outlines the change in the Bragg wavelength, 'ρ_α', 'α' and 'ξ' express the photoelastic, thermal expansion and thermo-optic coefficients of the fibre, respectively, '$\Delta\varepsilon$' defines the change of the mechanical strain and 'ΔT' depicts the temperature change.

For example, for a typical grating written in a conventional silica fibre and with $\lambda_B \approx 1{,}550$ nm, sensitivities to temperature and strain are, respectively, 10 pm/°C and 1.2 pm/µε.

The FBG shows some unique attributes, including

- Unlike fibre sensors produced in many type of optical fibres, which have bigger size than the original fibre and are often mechanically weaker, FBG sensor has exactly the same size as the original fibre and is of the same mechanical strength.
- The data are encoded onto the reflected light, hence FBG are unaffected by the drifts. The effect of strain and temperature is linear and additive to the FBG signal and does not require any additional calibration.
- By using the recently developed multiplexing techniques, various gratings can be written on a single optical fibre, leading thereby to a compact system with reduced cabling, enable to read simultaneously large numbers of sensors by using very few fibres.
- Both mechanical strain and temperature can be measured simultaneously with the same FBG. However, for industrial application, accurate measurement of temperature and strain should be done separately as following: one sensor sensitive to temperature and strain placed onto the structure under test and a second sensor close to the first one but isolated from the mechanical strain that allows it to respond only to the temperature.

These characteristics cited above, their flexibility and ability to manipulate any part of the optical transmission and/or reflection light, in addition to their relatively low cost, motivated the choice of FBG to be largely deployed for communications industry and advanced sensors systems.

1.1   Concept of fibre Bragg gratings interrogation

The FBG sensor is illuminated with a broad light spectrum, and then the reflected wavelength by the Bragg grating is measured. The occurred shift in the Bragg wavelength can be controlled by means of interferometry principle, which converts it into phase shift, hence easily detected by measuring the difference in the light-intensity following the variation of the light-path in the interferometer. Even though this is a costly technique (especially the equipment used), interferometry permits to reach a very high sensitivity.

As an alternative, another Bragg grating, namely the sloped optical filter, could be used as well to convert the wavelength shifts into intensity changes. As a matter of fact, this filter can be designed in such a way that a predefined pass and/or reject ratio changes with the wavelength; then any narrowband reflection from an individual grating can be straightforwardly resolved by comparing the passing and rejected intensities.

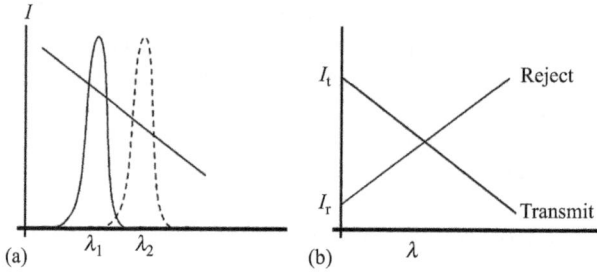

Figure 1.2   *Demodulating FBG with a passive filter: (a) the shifts of the wavelength are transformed into (b) intensity variations (I, I_t and I_r are the intensities of incident, transmitted and reflected light signals, respectively)*

Figure 1.2(a) shows the filter with a transmission spectrum. As the Bragg wavelength increments from 'λ_1' to 'λ_2', the transmitted light intensity 'I_t' declines, and the reflected (called also rejected) wavelength intensity 'I_r' rises subsequently. This is the straightforward and most cost-effective process of demodulating FBG; however, it is limited to only one grating at a time.

1.2   Wavelength-division multiplexing

For the wavelength-division multiplexing (WDM), various gratings are combined on a single optical fibre and are addressed at the same time, providing that, simultaneously, each grating will have a specific and different Bragg wavelength. Actually, this is accomplished either through a source of light with a broadband characteristics coupled to a spectrometer for the reflection detection, or by a light source having a tunable swept wavelength coupled to a simple photodiode as a detector. Figure 1.3 illustrates the WDM operating principle.

The scan generator adapts the light source by sweeping it onward/backward along its range, in such a way that the transmitted wavelength of light is known at any given instant. Once this transmitted wavelength concurs with the Bragg of a FBG, it is reflected back to the photodetector. The scan generator provides to the processor as well as timing signal helping to convert the light intensity with respect to the time. Additional processing to determine peaks of the reflected spectra and their relative positions and to convert them to strain and/or temperature is also accomplished.

Representative characteristics of the WDM system include

- Huge sensitivity and accuracy: Advanced WDM systems habitually carry out a wavelength resolution of about 1 pm, and long-term stability of 3.5 pm and above. Consequently, accuracy and strain sensitivity are of 4 and 0.8 με, respectively.
- Moderate speed: The scan rates are generally around 100 Hz, while certain sources can go up to 10 kHz; however, at these speeds, the data processing becomes challenging, especially when large numbers of sensors are employed.

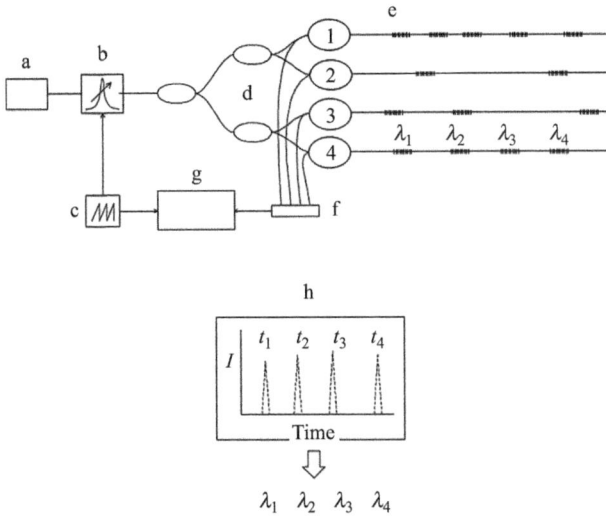

*Figure 1.3   Schematic and operating principle of WDM equipment: (a) light
source, (b) scanning filter, (c) scan generator, (d) coupler network
for channels 1–4, (e) FBG arrays, (f) photodetectors, (g) processor
and (h) time-varying output of the detector on channel 4, showing
times t_i converted into Bragg wavelengths λ_i*

- Flexibility: Theoretically, any number of FBG sensors can be used and placed
  on the optical fibre whenever they show different Bragg wavelengths. Weather
  they are located at 10 mm or 1 km apart, sensors are interrogated in the same
  manner. From a practical point of view, as the source of light and the FBG
  sensor reflections have both a determined widths of less than 0.5 and 50 nm,
  respectively, 100 sensors per channel are typically accommodated.
- Large size: Due to the need of an on-board processing power, the WDM control
  unit is of large size (comparable to a desktop computer). Compact units can
  obviously be accomplished at a high cost. Thus, WDM processors have been
  typically more appropriate to laboratory use or to some specific environments,
  especially in civil engineering and marine applications for which the size limitation is not an issue. WDM is less employed for spacecraft and aircraft vehicles.

1.3   Time-division multiplexing

A time-division multiplexing (TDM) system uses the echo time of a pulsed
broadband light wavelength source to reach the detector to identify the various
gratings. The echo obtained by a pulse from close neighbouring gratings is
obviously collected before those coming from faraway ones.

Figure 1.4 displays a schematic of an FBG array designed at various distances
'l' from the unit of the interrogation/detector. The light signal from the source (a)

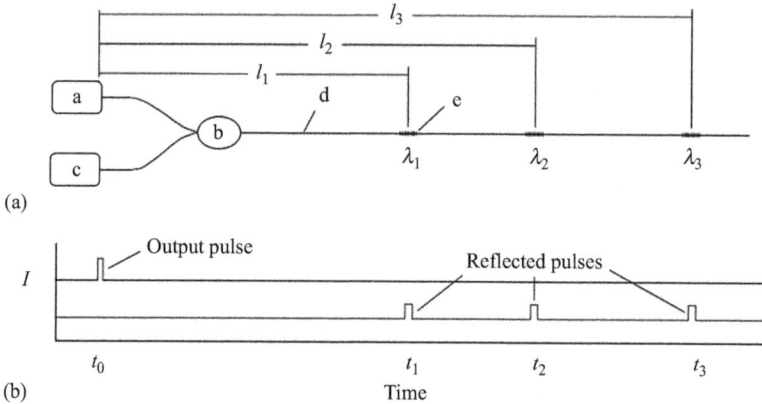

(a)

(b)   Time

Figure 1.4   Schematic of TDM system: (a) pulses from light source pass through coupler from 'a' to 'b', which is also connected to detector 'c', to fibre 'd' containing FBG 'e'. (b) Pulses emanating from source at time t_0 are reflected from FBG at positions l_1, l_2 and l_3, and return at t_1, t_2 and t_3

passes through the coupler shown in (b), which is linked to the light detector (c) and the optical fibre (d) that contains FBG array shown in (e). Pulses of light generated from the source at a time 't_0' are reflected from the FBG at distances 'l_1, l_2 and l_3' and then return at times 't_1, t_2 and t_3'. The required time 't_i' needed for a pulse echo from a given FBG located at 'l_i' is expressed by:

$$t_i = \frac{2l_i c}{n} \tag{1.3}$$

where 'c' is the speed of light (in vacuum) and 'n' is the RI of the fibre.

Once the accurate position of any grating in the array is established, we use a system of passive sloped filters to determine the corresponding wavelength of each echo pulse as it arrives.

The main attributes of a TDM system are as follows:

- Cost-effective: There is no need of particular specific tunable lasers or sophisticated filters; hence, the TDM system can be cheaper than the WDM one.
- Light and robust: For the TDM technology, all the processing is accomplished through a solid state electronic, without the need of moving parts; hence, the obtained system may be very compact and rugged, highly suitable for harsh environments conditions.
- Large sampling: The sampling rate is usually determined by the speed of the processing we employed and is not limited by the rate of the scan related to the source of light. Rates of sample in the range of kHz are easily obtainable.
- Sensor spacing: To be able to separate the echo-pulse temporally, one of the main technological restrictions for the TDM technology is the sensor spacing, meaning that they must be adequately distant from each other.

In summary, particular assets of using the FBG technology over other sort of FOS include their compact size, robustness, linearity, robust signal, simple multiplexing and linearity. This makes them ideal to be used as stable and very sensitive sensor for both temperature and mechanical strain. They can be used also into transducers for acceleration and pressure measurements. However, their dual sensitivity to temperature and mechanical strain at the same time can be seen as a technological limitation.

WDM interrogation concept is offering a high stability, flexibility, accuracy and high resolution in addressing multiple sensors and somewhat sampling with low rates. Although WDM are well matured, they are still somehow costly. Finally, although the TDM systems are limitedly employed, they offer rapid sampling at a lower cost. However, the sensor spacing is less flexible than that offered by WDM.

References

[1]   A. Orthonos and K. Kalli in *Fiber Bragg Gratings*, Artech House, Norwood, MA, USA, 1999.

[2]   R. Kashyap in *Fiber Bragg Gratings*, Academic Press, San Diego, CA, USA, 1999.

[3]   S. Nahar, C.J. Subhash, M. Vandana, *et al.*, *Current Science*, 2006, **9**, 2.

[4]   T. Sun, S. Pal, J. Mandal and K.T.V. Grattan in *Fibre Bragg Grating Fabrication using Fluoride Excimer Laser for Sensing and Communication Applications*, Central Laser Facility Annual Report 2001/2002, Rutherford Appleton Laboratory, Oxfordshire, UK, 2002.

[5]   K.O. Hill, D.C. Johnson, B.S. Kawasaki and R.I. MacDonald, *Journal of Applied Physics*, 1978, **49**, 5098.

[6]   Y.S. Kivshar and G. Agrawal in *Optical Solitons: From Fibers to Photonic Crystals*, Academic Press, London, UK, 2003.

[7]   K.W. Chow, I.M. Merhasin, B.A. Malomed, K. Nakkeeran, K. Senthilnathan and P.K.A. Wai, *Physical Review E*, 2008, **77**, 026602.

Chapter 2

Basic concepts, processes and material-based fibre optic sensors

Fibre optic sensors (FOS) technology has been under increasing development since more than 60 years and has led to the materialisation of many advanced devices, ranging from sensors of temperature, vibration, pressure and fibre optic gyroscopes to the highly sensitive chemical probes. This is mainly due to the various advantages FOS are offering, including their improved sensitivity – as compared to conventional sensing techniques – and geometric adaptability. This permits their configurations with different shapes. In addition, because the materials devices are dielectric, FOS can be used in harsh environment conditions, e.g. high voltage, high electromagnetic radiation, high temperature and even in corrosive milieu. Besides, FOS are also suitable for communications network and offer the remote-sensing ability. In this chapter, we review the basic concepts and processes related to FOS and some examples of doping materials used to improve the fibre sensitivity.

2.1 Fibre Bragg grating sensing basics

Presently, the research and development in the field of FOS-based devices has been extended to diverse technological application, including robotics, medical, environmental and telecommunications industries for space. Devices based on FOS were mainly developed to be able to asset a large set of physical properties, including mechanical strain, temperature, chemical transformation, pressure, rotation, displacement, radiation, vibrations and so on. As briefly discussed in Chapter 1, FOS are systems that can also operate in rough climate where traditional electronic and electrical devices meet complications. Unlike the conventional category of sensors, FOS displays various advantages, including the following:

- They are necessitating only small electrical cable size and are of small weight as they are non-electrical devices.
- Due to their small size, they can be employed to sensing into habitually inaccessible areas.
- They can be used for remote sensing.
- They are insensible to electromagnetic interference (EMI).
- They are chemically inert; they are not subject to contamination by their surrounding environment and are immune against corrosion.
- They grant high level of sensitivity and spectral resolution.

- They can provide sensitivity to multiple environmental parameters at the same time.
- They can be easily connected to any communication systems.

Figure 2.1 shows the main basic components of an FOS system, namely a light source, a transducer and a receiver.

For the optical source, lasers, diodes, and/or light-emitting diodes are mostly used.

An optical fibre (both single or multimode), doped fibres and bulk materials are often operating as the transducer (i.e. the sensor heart).

A photodetector is then used to identify the change in the wavelength (light signal) which is induced by the variation of the physical property under sensing (i.e. the perturbation of the sensed system). The photodetector is located at the output of the FOS system and allow the modulation of the polarisation, amplitude, phase and spectral signal of the light wave.

The optical modulation processes of the FOS devices implicated the following items:

- The amplitude change is proportionally associated to the reflection, scattering, transmission and absorption of the light wave signal. Presently, fibre Bragg grating (FBG) and long-period fibre gratings (LPFG) are used as the transducer in FOS network. The FOS factors that are usually modulated are the transmission, light wavelength, refractive index (RI) and reflexion. All these parameters are affected by the operation environment conditions.
- The variation in the phase is proportionally correlated to the frequency and/or wavelength change of the light. The variation in light wavelength is relative to the variation in the transmission, luminescence, reflection or absorption of the light, though the polarisation is associated to the strain birefringence.
- The transmission notion is commonly correlated to the rupture of an optical beam that is crossing along the optical fibre. The FOS that is operating through the optical reflection is using two ropes of fibres or a pair of single fibres. One rope of fibres relies on the light to a reflecting target, and the second ropes traps the reflected light and transmits it to the photodetector which detects the intensity change of the light, which in turn is the function of the environmental factors to sense.
- For the FOS that are based on microbending, only a modest amount of light is lost through the fibre walls when it is bent. Hence, under a physical perturbation

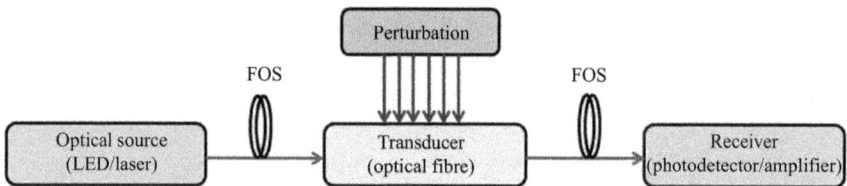

Figure 2.1   Main basic components of an FOS system, namely a light source, a transducer and a receiver

(such as under pressure), the FOS is mechanically bent and then the quantity of the received light varies proportionally corresponding to the perturbation parameter value.

Besides, the optical fibre can be chemically doped to change their optical parameters with a controlled way. Doing so, the measured optical parameter (absorption, reflection and so on) will be directly related to the characteristics of the incorporated chemical dopant. For example, novel bands will appear in the absorption spectrum and will be related to the given physical parameters, such as temperature.

Analogous to the absorption idea, luminescence is accomplished by chemically doping the fibre and/or some glass-based materials. In this category of sensors, the optical source generating light is employed to trigger a fluorescence signal, which in turn is proportionally affected by external environmental physical factors (e.g. radiation). In the same way, a variation in the luminescence level could be interpreted or transformed in a change of the colour which is the function of the level of the perturbing environment factor. Similarly, following a change of the environmental condition, a variation of the RI in the core of an optical fibre (e.g. fibre grating) occurs and could affect the optical frequency and, thereby, the amount of transmitted of reflected light received by the photodetector. In sum, the consolidation of all these concepts can be wisely employed with a multitude of mechanisms of light modulation to promote and complement the sensor function.

The FOS are classified in two main categories of sensors, namely the intensity-modulated ones and the phase-modulated ones:

- Intensity-modulated sensors: The principle of the intensity-modulated FOS is based on the change of the intensity of light which is subject of the environmental variation. This includes variation in the reflection, transmission and even the fibre microbending. Unlike the phase-modulated FOS, the intensity modulated ones usually require more amount of available light to operate accurately; consequently, they use a broad core multimode fibres or rope of fibres.
- Phase-modulated sensors: This category of sensor contrasts the phase of the optical signal in a sensing fibre by comparing it to a reference fibre in the interferometer. Commonly, these FOS use a source of light based on coherent laser and two fibres single mode. The optical signal is then separated (split) and implanted simultaneously into both the reference and sensing fibres. If the optical signal in the sensing fibre is subjected to the environmental perturbation (i.e. data to sense), a phase shift will occur as compared to that in the reference fibre and will be identified by the interferometer.

The phase-modulated sensors are very accurate compared to the intensity modulated ones. Typically, FOS can be conveniently categorised as a function of the way the optical fibre is employed. They can then be classified in terms of functionality as intrinsic and extrinsic sensors:

- Intrinsic FOS: These FOS use one optical fibre simultaneously as a sensitive material (i.e. sensor head) and as the milieu to guide the light that is subjected to the perturbation of the environment to sense. The intrinsic FOS operate *via* the direct modulation of the optical signal transported into the optical fibre. The parameter to sense should be strong enough to be able to modify the properties of the light that is carried through the fibre (Figure 2.2, upper panel).

These intrinsic sensors can employ interferometric structures, FBG, LPFG and chemically doped fibres that are specifically configured to be sensitive to particular perturbations.

- Extrinsic or hybrid FOS: In this kind of FOS, the role of the optical fibre is simply to transport the optical signal from and to a position where an optical sensor head is based. So, this sensor head is located externally relative to the optical fibre and is commonly fabricated with small and miniature optical components that are primarily constructed to modulate the light properties that are affected by the environmental conditions (i.e. to the physical perturbations).

In sum, in this particular configuration, first only one optical fibre will transmit the light energy to the sensor head. Second, this optical energy is accordingly regulated and is connected back through a second optical fibre that transports it along to the photodetector. This is the fundamental basis of the optical transmission sensor that is based on light-intensity modulation. Alternatively, the modulated optical signal could be also connected back into the same optical fibre through scattering and/or reflection (i.e. instead of transmission) and transported back to the photodetection system (Figure 2.2, lower panel). Hence, both intrinsic and extrinsic FOS systems operate similarly by the modulation of the wavelength, frequency, polarisation, phase and intensity of the transported optical signal.

Today, the FOS-based technology has become indispensable for process control in measurement systems, opening a large field of various applications, including medicine, telecommunications, computers, automotive industry, robotics, construction, agriculture, factory automation and space [1–8].

Novel technological challenges arose and are increasingly involving the control, the monitoring and the safety of processes. As a matter of fact, as an alternative to electrochemical-based technology, new FOS have been successfully

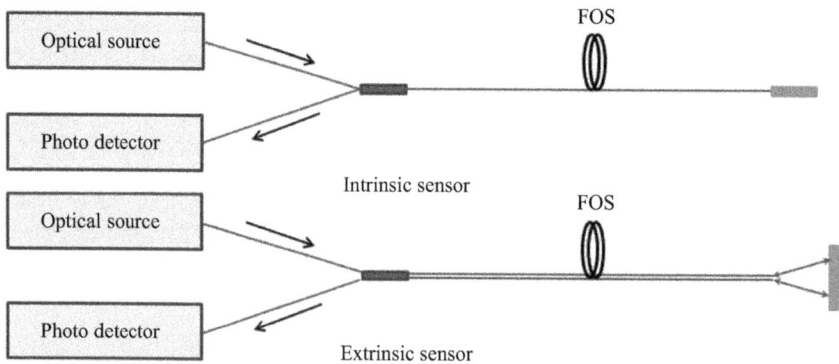

Figure 2.2 Schedule of an intrinsic and extrinsic FOS

implemented for the corrosion monitoring. In fact, corrosion in metallic structures is known to be a severe threat that implies costly maintenance and/or replacement processes, in addition to the interruption of the machine functioning that affects the global yield of the industry.

Generally, the rate of the corrosion in a metallic structure is assessed by its mass loss or through electrochemical processes. Alternatively, holographic interferometry [9,10] was established as one of the most efficient optical-based technology that is presently operating for monitoring the corrosion evolution. One of the most important restraints for this optical technique occurs when this monitoring needs to be conducted *in situ* at various laboratory-controlled conditions. In consequence, there is a need to explore new alternatives for such measurements. In fact, recently, FOS-systems-based intensity-modulated sensors were successfully investigated for the *in situ* corrosion monitoring [11–13] and paved the way to the further development of new FOS technologies that carry out *in situ* measurements.

2.2 Temperature compensation basics

In the following, we review the basics that depict the FOS-based FBG sensors. As mentioned earlier, in the FBG sensor that is made with silica, the Bragg shift is a function of strain and temperature, as expressed by the following equation:

$$\frac{\Delta\lambda_B}{\lambda_B} = (1 - \rho_e)\varepsilon_z \tag{2.1}$$

where $\Delta\lambda_B$ is the change in the Bragg wavelength, ρ_e is the photoelastic coefficient and ε_z is the longitudinal strain of the grating.

ρ_e expresses the variation of the index of refraction with respect to the strain:

$$\frac{\Delta\lambda_B}{\lambda_B} = (1 - \rho_e)\varepsilon_z + (\alpha + \eta)\Delta T \tag{2.2}$$

where α is the thermal expansion of silica, λ_B is the Bragg wavelength, η is the thermo-optic coefficient expressing the dependence of the RI with respect to the temperature (i.e. $\eta = dn/dT$). e: dn/dT. ΔT is the variation of the temperature.

For a FBG designed on a conventional optical fibre based on silica and doped core with Ge, $\rho_e = 0.22$, $\alpha = 0.55 \times 10^{-6}/°C$ and $\eta = 8.6 \times 10^{-6}/°C$.

Hence, at the wavelength of 1,550 nm, after substituting the constants in (2.2), the sensitivity of the FBG (i.e. the grating) to temperature and strain is:

$$\frac{\Delta\lambda_B}{\lambda_B} = 14.18 \text{ pm}/°C \tag{2.3}$$

and

$$\frac{\Delta\lambda_B}{\lambda_B} = 1.2 \text{ pm}/\mu\varepsilon \tag{2.4}$$

where '$\mu\varepsilon$' is the axial strain of the FBG.

Nonetheless, these theoretical parameters are not absolute as each FBG sensor is somehow unique. In fact, even fabricated from the same batch, the FBG will necessarily show different sensitivities.

Moreover, by taking into account only the observed $\Delta\lambda_B$ parameter, one cannot certify if the FBG shift (i.e. displacement) in a consequence of the strain or temperature, or both simultaneously. Hence, if we target to extract only the temperature effect, the FBG sensor needs to be protected against the strain. This could be carried out by simply introducing the FBG into small-bore rigid tubing. Nevertheless, if we target to record solely the strain, the scenario is much more complicated, due to the difficulty to stop the change of local temperature reaching the FBG sensor. Alternately, we can compensate this temperature variation by measuring the local temperature through a thermistor and then calculate its effect on the Bragg wavelength shift. This shift will be extracted from the total one as measured. It is worth noting here that this approach is possible only if we can access to measure electrically the temperature, which is not evident especially if the surrounding environment is characterised by EMI and/or high-voltage conditions.

Another alternative is to use a second FBG sensor (say FBG2) on the same fibre, which is protected against the strain as mentioned above, but put at the same temperature of the first sensor (say FBG1). The two FBG located in the same optical fibre will furnish two different Bragg reflections: one dependent on the strain and temperature and a second one which depends only on the temperature, for compensation.

2.3   Calibration of fibre Bragg grating sensors with temperature and evaluation of the uncertainty

In reality, (2.2) is not an accurate standard for the FBG attitude operating under temperature change, and hence every FBG sensor has to be separately calibrated to sense precisely the temperature through the Bragg wavelength.

In this section, we report on an avant-garde research conducted by Werneck *et al.* [14] to align a fibre optic chain formed of five FBG sensors. The five FBG sensors were exposed to a temperature gradient of 65°C (variation from 20°C to 85°C) to verify and subsequently quantify the parameters of (2.2).

The number of FBG necessary to sense the temperature in different Bragg wavelengths is not limited, as long as each FBG spectrum is not overlapping with each other during its shift (displacement) with respect to the temperature variation. One solution is to distribute these sensors along the usable range of FBG interrogators. In this particular study, the wavelength range was over 40 nm, namely 1,530–1,570 nm.

Figure 2.3 shows the experimental setup employed to calibrate the FBG sensors. The dotted square expressed the optical system constituted of a commercial Bragg meter tool. This tool includes a broadband source of an amplified spontaneous emission used to irradiate the FBG through port 1 of the optical circulator. Then, the FBG reflection spectrum backs *via* port 2 and is then conducted through port 3 to an embedded optical spectral analyser. At this level, the detected reflected spectrum is measured. All controls and data can be accessed by a computer connected to the USB port of the interrogator.

Figure 2.4 displays the spectra of five FBG sensors exposed to a temperature variation from 20°C to 85°C [14].

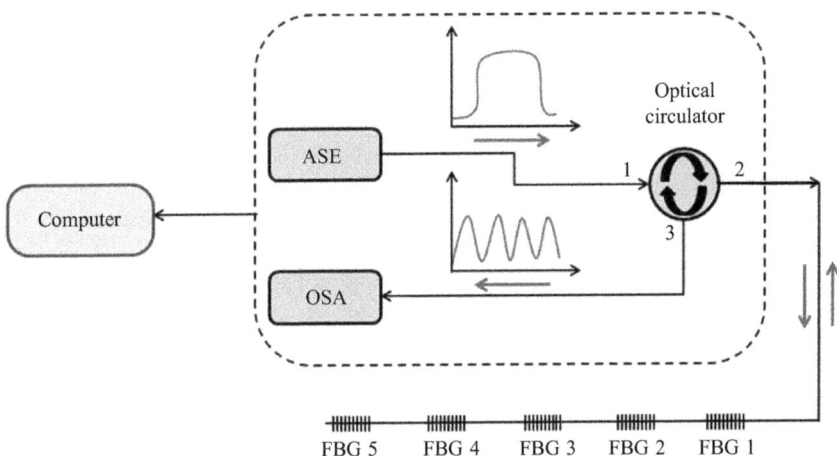

Figure 2.3   Schematic diagram of the measurement technique employing five FBG sensors exposed to temperature variation between 20°C and 85°C

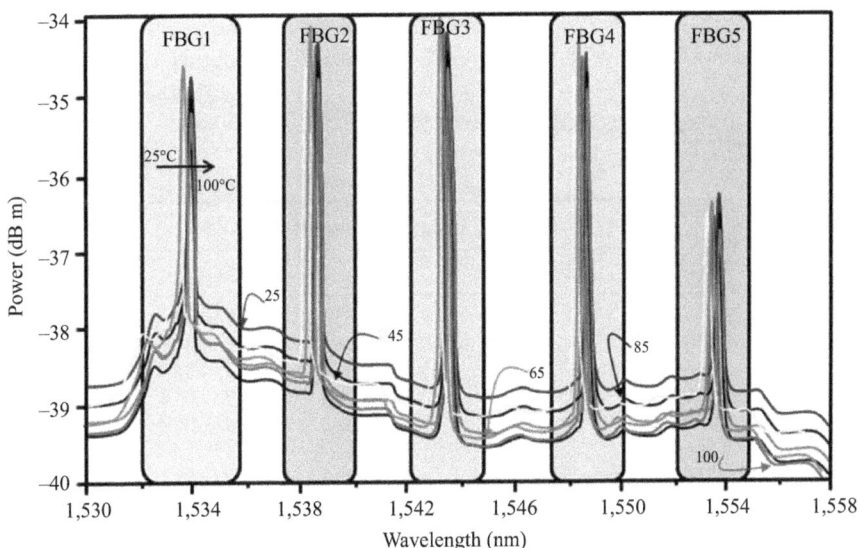

Figure 2.4   Five FBG sensors spectra recorded at temperatures from 25°C to 100°C

The method Werneck *et al.* followed to calibrate the sensors is based on the following:

- First, all FBG sensors were immersed simultaneously into a batch with a controlled temperature. The Bragg wavelength displacements were then registered together with the temperature as displayed by thermometer (a National Institute of Standards and Technology-traceable, TD 990, Thermolink, 0.1°C resolution and ±1°C accuracy).
- Five groups of measurements were carried out for each sensor in the range of 20°C–85°C.

Table 2.1 shows the Bragg displacement for each temperature and for each FBG sensor. As we can see, it is easily possible to estimate the sensitivity of each FBG sensor, and the accuracy of the measurement chain. The plot in Figure 2.5 was displayed from the data shown of Table 2.1.

Table 2.2 summarises the following calibration parameters:

- The sensitivities (both experimental and theoretical)
- The coefficients of correlation of the plots fittings
- The root mean square error (RMSE)
- The maximum residual errors.

Figure 2.6 displays the analysis of the error regarding the first sensor: FBG1. The maximum residual error was roughly 0.004 at 30°C. All the other temperature error measurements for the other four sensors were within the range of ±0.007°C.

All R^2 were found to be very close to the unity, and the errors as such were found to be much smaller than 1°C.

Note that these errors originate from the uncertainties of the interrogation system (±1 pm) and from those of the employed thermometer. By adopting 13 pm/°C as the

Table 2.1   Average Bragg centre wavelength of each FBG under temperature variation

T (°C)	Average Bragg wavelength peak (nm)				
	FBG1	**FBG2**	**FBG3**	**FBG4**	**FBG5**
25	1,536.001	1,540.928	1,545.833	1,550.819	1,555.723
30	1,536.067	1,540.995	1,545.903	1,550.888	1,555.793
35	1,536.128	1,541.059	1,545.971	1,550.959	1,555.865
40	1,536.183	1,541.116	1,546.030	1,551.015	1,555.922
45	1,536.245	1,541.177	1,546.092	1,551.083	1,555.987
50	1,536.310	1,541.247	1,546.163	1,551.147	1,556.055
55	1,536.368	1,541.308	1,546.224	1,551.213	1,556.119
60	1,536.427	1,541.368	1,546.284	1,551.273	1,556.178
65	1,536.497	1,541.440	1,546.358	1,551.345	1,556.253
70	1,536.557	1,541.501	1,546.420	1,551.408	1,556.317
75	1,536.618	1,541.566	1,546.484	1,551.469	1,556.383
80	1,536.680	1,541.627	1,546.549	1,551.539	1,556.447

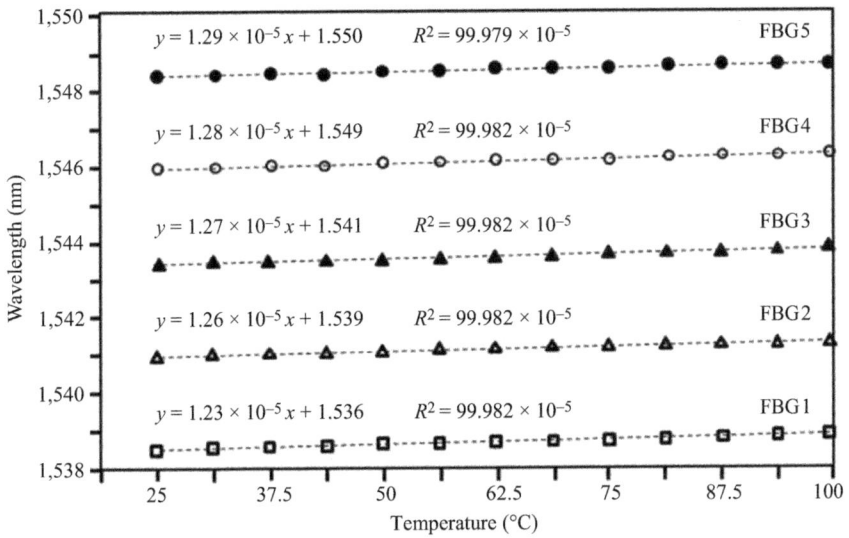

Figure 2.5 Experimental wavelength shift measured as a function of temperature for each of the five FBG sensors

Table 2.2 Calibration parameters

FBG	Theoretical sensitivity (pm/$^\circ$C)	Measured sensitivity (pm/$^\circ$C)	Correlation coefficient (R^2)	RMSE ($^\circ$C)	Residual error ($^\circ$C)
1	14.05	12.31	0.99982	0.00311	0.003
2	14.10	12.71	0.99981	0.00328	0.005
3	14.14	12.95	0.99982	0.00327	0.005
4	14.19	12.99	0.99993	0.00320	0.006
5	14.23	13.10	0.99978	0.00363	0.007

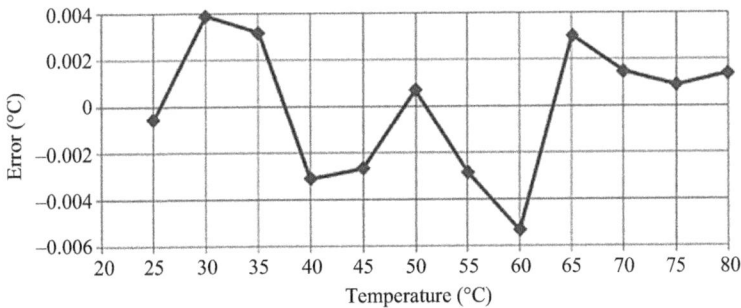

Figure 2.6 Error analyses for FBG1 sensor

FBG average sensitivity (Table 2.2), 1 pm in error translates a temperature fluctuation of about 0.08°C, which is still much lower than that produced by the classical thermometer.

From data shown in Table 2.2, we can see that the theoretical sensitivities as anticipated by (2.5) are quite different from those acquired by the calibration experiment:

$$\frac{\Delta \lambda_B}{\lambda_B} = 2n_{eff}\frac{\partial \Lambda}{\partial T} + 2\Lambda\frac{\partial n_{eff}}{\partial T} \tag{2.5}$$

where λ_B of an FBG which is a function of the effective RI of the fibre n_{eff}, T is the temperature and the Λ is the grating period.

Also, based on the Bragg's law,

$$2d \sin \theta = n\lambda \tag{2.6}$$

where θ is the incident angle, λ is the wavelength, d is radiation dose and n is an integer.

As detailed in Chapter 5, the fabrication of the FBG is not an automatic process where the radiation time/intensity for each FBG writing is not constant. In fact, during the fabrication process, the ultraviolet (UV) laser is disconnected by the operant only when the reflection of the Bragg grating turns up clearly above the desired level (threshold).

The UV radiation affects and modifies the RI of the core of the optical fibre and modifies by the same the values of η in each FBG sensor, individually and differently. Gwandu et al. [15] were the first confirming such an effect and have successfully validated a process for controlling and varying the temperature responsivity of each FBG by simply increasing the UV exposure time.

From the data in Table 2.2, for each FBG sensor, it was possible to come out with a linear correlation between the wavelength and temperature. The value of $R^2 = 0.99996$ demonstrates clearly the linear proportionality and the accuracy of FBG sensors for temperature measurements.

2.4   Photosensitivity in optical fibres

Conceptually, photosensitivity is the threshold of received photons starting from which the irradiated object reacts. For the glass material, the photosensitivity was first demonstrated by Hill et al. [16], in 1978, at the Communications Research Center in Canada.

This non-linear effect freshly established in optical fibres was first called fibre photosensitivity. In 1988, Meltz et al. [17] suggested a decade later a model for fibre photosensitivity. Their model was primarily motivated by the fact that fibre photosensitivity was identified only in Ge-doped silica fibres, and thus, they based this model on the synthesis conditions of the doping. Indeed, when Ge-doped silica fibres are elaborated by modified chemical vapour deposition process at high temperatures, germanium dioxide (GeO_2) and silicone dioxide (SiO_2) associate together in the form of gases to fabricate the fibre core. During this step, there is a high probability that products like Ge–Ge, O–Ge–O and Ge–Si could be constituted and might create defects in the fibre lattice. These by-products are named in the relevant literature as 'wrong bonds'.

The fibre thus shows huge absorption located at 245 nm that is corresponding to the presence of these defects. Then, once these defects are excited with UV laser, some absorption lines occur and the RI increases at these bands. However, there is no consensus and no single model that is explaining the experimental results presented in the literature regarding the origins of the photosensitivity and the variation of the RI. These phenomena are not yet fully understood. Consequently, it is clear that photosensitivity is a function of various photochemical, photomechanical and thermochemical mechanisms [5].

One of the theoretical models that show the best agreement with the experimental data seems to be the compaction (or compression) type based on the Lorentz–Lorenz law. This model suggests that the RI increases with respect to the material compression. This concept was overtaken, and these equations were compiled into the FBG interrogation system software that gives the temperature of each sensor.

Finally, as shown in Figure 2.7, it is feasible to draw the temperature of calibration as a function of the experimentally measured temperature [18], by using UV laser excitation to create a thermally reversible and linear compression in the amorphous SiO$_2$. A compiled incident irradiated dose of 2 kJ/cm^2 would be sufficient to create an irreversible densification and produces a photoetching. In addition, the results obtained by laser compaction or densification were found to be in good agreement with those obtained using hydrostatic pressure. Consequently,

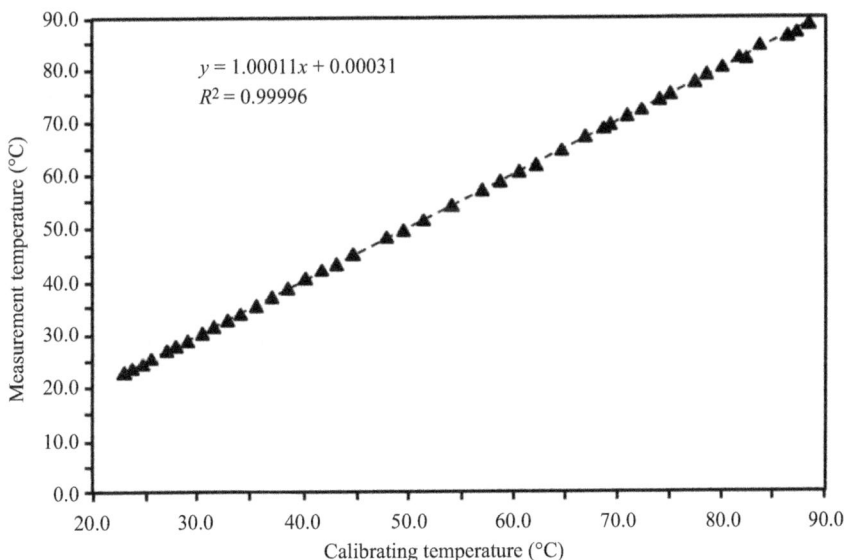

Figure 2.7 Temperature calibration responses of the FBG1 sensor

a linear trend of the evolution of the RI with respect to the variation of the Bragg period was observed, corroborant the predictions of the Lorentz–Lorenz law.

It is known that when an optically transparent material is mechanically compressed, two main phenomena that affect the RI were raised, namely (i) the increase of the RI owing to the increase of the material density and (ii) the photoelastic effect. The latter has negative impact for many optical media. Nonetheless, the compression could produce an increase of the RI in the illuminated parts of the FBG, which is a much stronger effect than the photoelastic one.

With the evidence that UV irradiation triggers an increase of the RI, we can presently notice physically the FBG impact. In fact, Figure 2.8 displays the change of the fibre core's RI (n_{core}) along the length of the z-axis. The effective RI of the fibre core is the average RI of the irradiated portion of the fibre core and its value is around 1.45. Moreover, due to the UV radiation, the change of the RI is about Δn, which is the amplitude of square diffraction pattern and is equal to 10^{-4}. The grating period 'Λ' is the same as the interference pattern is about 500 nm, and the FBG length (L_{FBG}) is around 10 mm.

From the other side, the variation of the interference pattern is not occurring as a square wave but rather as an approximate sinusoidal waveform. This type of variation will trigger an RI change on the fibre of the same form, as it is shown in Figure 2.9.

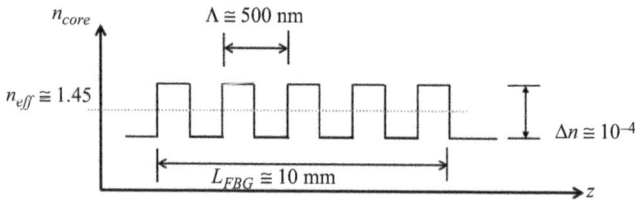

Figure 2.8   *Variation of the RI of the fibre core (n_{core}), n_{eff} is the average RI of the original fibre core along the length of the fibre L_{FBG} (z-axis) resulting from a square wave diffraction pattern*

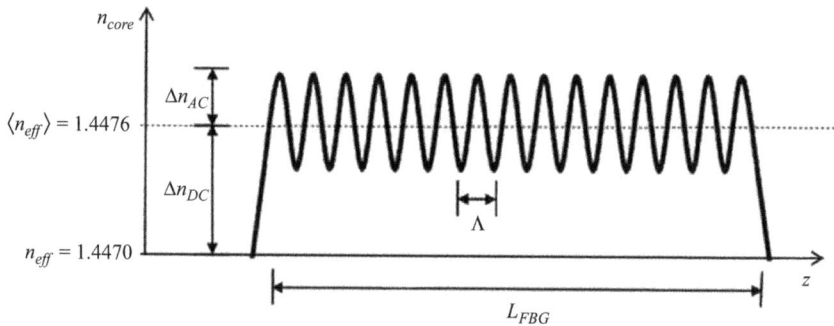

Figure 2.9   *Simulation of the variation of the RI of the fibre core (n_{core}) along the length of the fibre (z-axis) resultant of a sinusoidal diffraction pattern*

In this figure, n_{eff} is the average RI of the original fibre core along the FBG region resulting from a square wave diffraction pattern, and $\langle n_{eff} \rangle$ is the average RI of the original fibre core, along the FBG region, resulting from a sinusoidal diffraction pattern. Δn_{DC} is the average amount of RI that is enhanced by the UV dose, and Δn_{AC} is the half of the total RI change in the FBG.

Equation (2.7) describes the theoretical model of the RI in the FBG area, with respect to the radiation dose d of the UV light and the distance along the axis of the fibre [19]:

$$n_{eff}(z,d) = n_{eff} + \left[\Delta n_{DC}(d) + \Delta n_{AC}(d)\sin\left(\frac{2\pi}{\Lambda}\right)(z) \right] \tag{2.7}$$

Both Δn_{DC} and Δn_{AC} were found to increase proportionally with the UV radiation dose. However, as the UV radiation is never zero along the diffraction region, all FBG lengths undergo an increase of the RI.

We can then rewrite (1.1) by using the average RI in the FBG area as following:

$$\lambda_B = 2\langle n_{eff} \rangle \Lambda \tag{2.8}$$

Prior to the FBG inscription (the set-up used for is detailed in Chapter 5), it is necessary to first calculate the angle φ for the desired Bragg wavelength. Second, the operator will monitor the reflection spectrum until the reflectivity matches the aimed value. However, during the UV irradiation, as $\langle n_{eff} \rangle$ increases, λ_B will increase as well proportionally. Figure 2.10 displays the typical progression of an FBG reflection spectrum as a function of the UV dose.

The dotted line in Figure 2.10 shows the direction of the Bragg wavelength as the UV dose increases.

Therefore, two relationships can be obtained from this plot:

$$Reflectivity = f(number\ of\ shots\ or\ irradiation\ time) \tag{2.9}$$

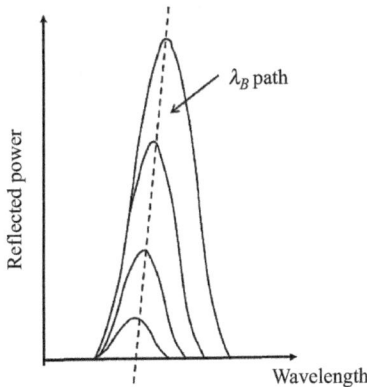

Figure 2.10   Behaviour of a FBG reflection spectrum with respect to the UV irradiation dose

and

$$\lambda_B = g(number\ of\ shots\ or\ irradiation\ time) \qquad (2.10)$$

$f(*)$ and $g(*)$ are arbitrary functions that are defined by plot fitting.

During the FBG writing using pulsed UV laser, both the reflectivity and the Bragg wavelength enhance with respect to the number of laser pulses. Every laser pulse produces a given dose of UV light energy. The total energy is then integrated over the number of laser pulses which is generating the stress inside the fibre core. Equations (2.9) and (2.10) are of tremendous importance during the FBG design.

On the other hand, the increase of the reflectivity as a function of the number of laser shots and/or UV dose is limited by an upper threshold, above which the reflectivity starts decreasing. This is the result of competing effects between the RI and the photoelastic effect. Indeed, for many optical media, while the RI increases because of the increase of the density of the material, the photoelastic effect is rather negative. Hence, above the threshold plateau, the photoelastic effect becomes higher than that due to the effect of the material density increase, because the latter saturates while the photoelastic effect does not.

As a matter of fact, this threshold is of 500 mJ/cm^2 for Type I gratings FBG. Moreover, during its formation, λ_B undergoes a redshift (Figure 2.10), while in Type IIA formation, it experiences rather a blueshift one. Above this threshold, the FBG starts to be effaced until it fully disappears.

It is known that in the optical fibre-based germanosilacate, the Bragg grating wavelength decreases with respect to the temperature. In Type I gratings, FBG are rather stable up to 300°C, and up 500°C in Type IIA gratings [5]. It is also recommended that after the gratings fabrication, especially in Type I, FBG are subjected to an annealing process at temperature above their operating one to produce an accelerating ageing and stabilise the FBG.

From industrial perspective, telecommunication fibres usually contain a concentration of about 3.5% of germania (GeO) doping but are feebly responding to a UV radiation ($\Delta n = 10^{-5}$) and are showing low photoreflectivity (i.e. not suitable for FBG application). In fact, only a doping concentration of 5% and above will lead to a useful amount of photosensitivity. However, the fabrication cost increases with doping concentration in the fibre. Finally, GeO-doped fibres are often used for long-distance sensing, e.g., to remotely monitor environmental parameters that are located many kilometres away from the interrogation unit.

Some fibres are doped with boron co-doped GeO. They show an enhanced sensitivity but enhanced losses as well, which make them unsuitable for long-distance sensing and are often limited to few metres only.

The hydrogen loading and diffusion into the fibre core is one of the best process to enhance the photosensitivity. The mechanism behind that is believed to be due to the reaction of hydrogen H_2 with GeO. In heavily doped fibres, there is a high concentration of Ge–O–Ge bonds which react with hydrogen H_2 to form Ge–OH molecules that absorb preferentially UV radiation and increase thereby the internal stress into the core of the optical fibre [20].

The diffusion of hydrogen is achieved by putting the fibre inside a high pressure and tight enclosure hydrogen box. Typical pressures of 20–750 atm are commonly used, where 150 atm is the most employed. In addition to increase the photosensitivity of the fibre, hydrogen loading grants the writing of FBG sensors in both germane silicate and Ge-free fibres.

The fibre becomes very soft when removed from the hydrogen-loaded high-pressure vessel; however, by heating it up for just few minutes, the hydrogen diffuses out.

The loading of hydrogen can be achieved as well through the flame brushing process, which consists to burn the optical fibre by a hydrogen oxygen flame, at temperatures reaching 1,700°C, for 20 min. At this level of high temperature, the hydrogen surplus in this mixture diffuses into the fibre. The asset of this frame brushing process is the possibility to elaborate conventional telecommunications fibres, while the drawback is the fact that, in case of fibre-based acrylate buffer, the flame will burn an area larger than that of writing the FBG itself, which requires a posterior recoating.

2.5 Type of gratings

The term 'type' invokes the mechanism of the photosensitivity with which the grating fringes are formed in the optical fibre. The different processes in writing these grating fringes lead to different physical properties of the produced FBG. The temperature sensing and the capacity of the FBG to resist to elevated temperatures are the two most fabrication-dependent parameters. These writing approaches are summarised as follows:

- Type I gratings, also called the standard type: Manufactured in optical fibre of all types, hydrogenated and non-hydrogenated. The Type I gratings is also known as standard gratings. Commonly, the R-reflection spectra of this standard Type I gratings is straightforwardly equal to $1-T$ (T being the transmission spectra), meaning that transmission and reflection spectra are complementary, and there is negligible or almost no loss of light occurring either by absorption and/or reflection into the cladding. This common Type I gratings is typical and the most used one among all grating types. It is the only one type of grating available off-the-shelf during the fabrication (writing) process.

- Type IA gratings: In 2001, during research experiments conducted to determine the hydrogen loading effects on the writing process of Type IIA gratings in germanosilicate optical fibre, we observed for the first time the Type IA gratings [21]. Unlike a blueshift that is typically observed in the gratings Bragg wavelength, a large redshift is rather observed in this Type IA. Subsequent work has demonstrated that the blueshift (i.e. increase in the Bragg wavelength) occurred once an initial Type I grating had reached peak reflectivity and started to diminish. For this reason, Type IA gratings was designated and labelled as 'a regenerated grating'. Moreover, the temperature coefficient of the Type IA gratings has found to be lower than that of a standard grating that is written under similar conditions.

- Type IIA gratings: These gratings are written as the negative part of the induced index variation. It is commonly affiliated with gradual relaxation of induced stress along the axis and/or at the interface level [22]. In the work conducted by Xie *et al.* [23], the presence of another kind of gratings having the same thermal stability as for the Type II gratings was demonstrated. This new type of gratings has found to show a negative variation in the mean index of the fibre and was labelled Type IIA. The gratings were written in germanosilicate fibres by means of a pulsed xenon monochloride laser. First laser irradiation will form a standard (Type I) grating within the optical fibre that undergoes a small redshift prior to be subsequently erased. Further laser exposure has shown that a grating is reformed which undergo a steady blueshift whilst growing in strength [23,24]. The main difference between the writing of Type IA and IIA gratings is that Type IA is formed in hydrogenated fibres only, while Type IIA is developed in non-hydrogenated ones [25,26].

References

[1]   B. Culshaw, *Journal of Lightwave Technology*, 2004, **22**, 1, 39.

[2]   F.T.S. Yu, S. Yin and P.B. Ruffin in *Fibre Optic Sensors*, 2nd Edition, CRC Press Taylor & Francis, Boca Raton, FL, USA, 2008.

[3]   D.A. Krohn in *Fibre Optic Sensors, Fundamental and Applications*, 3rd Edition, Instrument Society of America, Research Triangle Park, NC, USA, 1999.

[4]   J.M. Lopez-Higuera in *Handbook of Optical Fibre Sensing Technology*, 1st Edition, John Wiley & Sons, Hoboken, NJ, USA, 2002.

[5]   A. Othonos and K. Kalli in *Fibre Bragg Gratings: Fundamentals and Applications in Telecommunications and Sensing*, Artech House Publishers, Norwood, MA, USA, 1999.

[6]   V.K. Rai and S.B. Rai, *Applied Physics B: Lasers and Optics*, 2007, **87**, 2, 323.

[7]   V.K. Rai, *Applied Physics B: Lasers and Optics*, 2007, **88**, 2, 297.

[8]   E. Udd in *Fibre Optic Sensors: An Introduction for Engineers and Scientists*, Wiley Interscience, Hoboken, NJ, USA, 1991.

[9]   K. Habib, *Optics and Laser in Engineering*, 1993, **18**, 2, 115.

[10] K. Habib, *Optics and Laser in Engineering*, 1995, **23**, 1, 65.

[11] J. Castrellon-Uribe, C. Cuevas-Arteaga and A. Trujillo-Estrada, *Optics and Lasers in Engineering*, 2008, **46**, 6, 469.

[12] S. Dong, Y. Liao and Q. Tian, *Applied Optics*, 2005, **44**, 27, 5773.

[13] S. Dong, Y. Liao and Q. Tian, *Applied Optics*, 2005, **44**, 30, 6334.

[14] M.M. Werneck, R.C. Allil and B.A. Ribeiro, *IET Science, Measurement & Technology*, 2013, **7**, 1, 59.

[15] B.A.L. Gwandu and W. Zhang, *IEEE Sensors*, 2004, **3**, 1430.

[16] K.O. Hill, Y. Fujii, D.C. Johnson and B.S. Kawasaki, *Applied Physics Letters*, 1978, **32**, 10, 647.

[17] G. Meltz, W.W. Morey and W.H. Glenn, *Optics Letters*, 1989, **14**, 15, 823.

[18] K. Tanimura, T. Tanaka and N. Itoh, *Physical Review Letters*, 1983, **51**, 423.

[19] F. Jülich and J. Roths *Proceedings of the OPTO 2009 and IRS² 2009*, Nuremberg, Germany, 2009, 119.

[20] J. Albert, M. Fokine and W. Margulis, *Optics Letters*, 2002, **27**, 809.

[21] Y. Liu in *Advanced Fibre Gratings and their Application*, Aston University, Birmingham, UK, 2001, p. 23. [PhD. Thesis].

[22] J. Canning, *Lasers and Photonics Reviews*, 2008, **2**, 4, 275.

[23] W.X. Xie, P. Niay, P. Bernage, *et al.*, *Optics Communications*, 1993, **104**, 1–3, 185.

[24] P. Niay, P. Bernage, S. Legoubin, *et al.*, *Optics Communications*, 1994, **113**, 1–3, 176.

[25] A.G. Simpson in *Optical Fibre Sensors and Their Interrogation*, Aston University, Birmingham, UK, 2005, p. 57. [PhD. Thesis].

[26] A. Rahman and S. Asokan, *International Journal of Smart Sensing and Intelligent Systems*, 2010, **3**, 1, 110.

Chapter 3

Harsh environment fibre Bragg grating sensing

The most ubiquitous and dominant current technologies employed for temperature measurements are those based on electrical sensing, including thermocouples, thermistors and resistance temperature detectors. For standard temperature-sensing applications, these electrical sensors are definitely the best solution from economical point of view; however, for challenging applications in harsh environmental conditions, optical sensing can offer an imperative alternative. Harsh environments conditions, long-term distributed and long-term implementations are a few typical examples where the characteristics of an optical-sensing system is advantageous and can make an effective and more relevant solution as compared to conventional electrical sensors system. In this chapter, we will briefly focus on the temperature measurement applications of fibre Bragg grating (FBG) sensors, together with their behaviour under high mechanical compression and gamma irradiation. Examples of these applications will be detailed later on, in Chapters 6–8, including our most recent experimental data.

3.1 Introduction and motivation

Optical fibres find use in other fields than telecommunications. A very promising and expanding field is the fibre optic sensing. In comparison with standard electrical sensor systems, the fibre optic alternative presents many relevant advantages:

- First, the optical fibre accesses to the vast number of multiplexed points, which can substitute several thousand of electrical point detectors.
- Second, as the data are carried out optically through light propagation, the electromagnetic interference (EMI) and/or jammer are harmless.
- Third, as the fibre is galvanically isolated from the measured entity, a direct contact with the sensed object is hence possible.

From practical point of view, it is often desirable to measure temperature distributions in boilers, turbine engines, furnaces, chemical reactors and others. Distributed high-temperature sensing can enable real-time monitoring of combustion processes, thus helping to increase the fuel efficiency and reducing, for e.g., the nitrogen dioxide pollution, which is a big potential cause of cancer.

However, the extreme conditions of temperature encountered in such applications make measurement difficult. Thermocouples are frequently employed to measure temperature, and while thermocouples can generally tolerate fairly high temperature and pressure conditions, thermocouples are point sensors that can only sense the temperature at one particular location. In order to access to a precise three-dimensional spatial information of a temperature distribution, a number of thermocouples are needed which will be much more costly than optical fibres. Furthermore, thermocouples can be very sensitive to external electromagnetic fields, which deteriorate their accuracy and preclude their use in a number of applications. The industry has attempted to utilise various optical sensors such as blackbody radiation-based sensors which employ a fibre optic to transmit a radiation signal. As for the thermocouples, these sensors are based on point-by-point sensing and are then not suitable to furnish the temperature distribution data unless a large network of sensing devices is employed. Other optical sensors rely upon interferometry. But these systems are fairly complicated, susceptible to mechanical vibrations and hard to realise distributed-sensing networks.

One of major advantages of grating-based fibre optic sensor(s) (FOS) is that they can be multiplexed, which enables multipoint sensing along one sensing fibre. Each grating is inscribed at different locations on the fibre with different grating periods. Then, the signals coming from each sensor are encoded at different positions in the wavelength domain. The physical quantity we want to measure, temperature in this case, makes the spectral peak shift in the wavelength range. By measuring the shift, temperature sensing can be realised.

By writing a group of gratings in the same optical fibre, distributed sensing could be accomplished. This novel process offers an enhanced sensitivity, great constancy and reliability, and distributed-sensing capability, which is highly suitable for high-temperature distributed sensing. So far, various kinds of optical fibre grating-based sensors have been developed and can be categorised into two main types, namely (i) FBG and (ii) long-period fibre gratings (LPFGs). The latter has also been successfully determined to have high sensitivity to the change of refractive index (RI) of the sensed environment (called here ambient media). However, the multiple resonance peaks and broad resonance bandwidth features of the LPFG lead the spectra to overlap, limiting thereby their precision and reducing their multiplexing capabilities. Moreover, their long length will limit their application as point sensor devices. On the other hand, owing to their huge potential, including narrow bandwidth, high sensitivity, great resolution, wavelength multiplexing, and distributed-sensing capabilities [1], the FBG were extensively investigated and matured. In the following section, we will focus on the temperature measurement applications of FBG sensors.

3.2 Temperature sensing principle

The continual growth of the FOS developed to operate in extreme conditions and harsh environments (such as for temperatures up to 1,000°C, nuclear radiation and so on) led to an increasingly new applications. Among FOS, FBG sensors are largely employed for structural health monitoring and ambient sensing. Their main assets in comparison with other FOS technologies include their multiplexing capability, the measurement of reflected light and the wavelength-encoded sensing ability [2].

FBG are established by irradiating the optical fibre with an intense ultraviolet (UV) laser-interference pattern that forms a periodic RI modulation. However, this RI modulation is not permanent and depends on the fibre type. For example, for high-temperature sensing operations, this modulation decreases (staring from 500°C) until total extinction which happens typically around 600°C–700°C [3,4]. Several approaches and techniques have recently been developed to improve the FBG temperature stability. One of the most encouraging processes requires the use of chemical composition gratings (CCGs) [4–7].

CCGs are constructed from hydrogen-loaded optical fibres that are subsequently subjected to an annealing treatment at high temperature. This process alters the FBG RI modulation by a chemical structure that is more stable. During the annealing process, the original FBG is completely eliminated and a new RI modulation is produced in the areas that were antecedently UV irradiated. These new gratings are known as regenerated FBG. CCG were successfully developed using two different types of optical fibres, namely the standard telecommunications Ge-doped fibres and the Ge–B co-doped photosensitive fibres.

The huge sensitivity of cladding modes to surrounding the values of the RI makes FBG also suitable for chemical sensing [1–9]. Figure 3.1 presents a typical example of the shift of the grating resonant wavelength with respect to the increasing temperature. Figure 3.2(a) and (b) displays a representative case of the transmission spectrum and the wavelength shift of various FBG sensors as function of temperature. Figure 3.2(c) shows the resonance wavelength shift as the environmental RI changes. Similar to the temperature sensitivity, the refractive sensitivity of the FBG is also mode dependent.

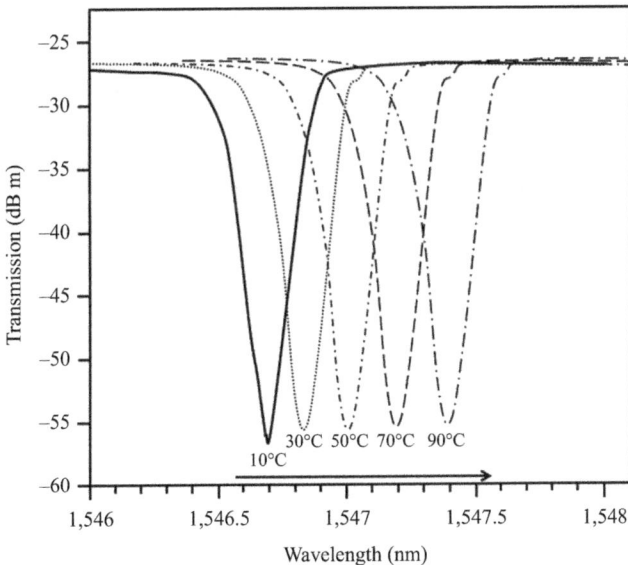

Figure 3.1   Typical example of measured shift in the grating spectrum with respect to increasing temperature

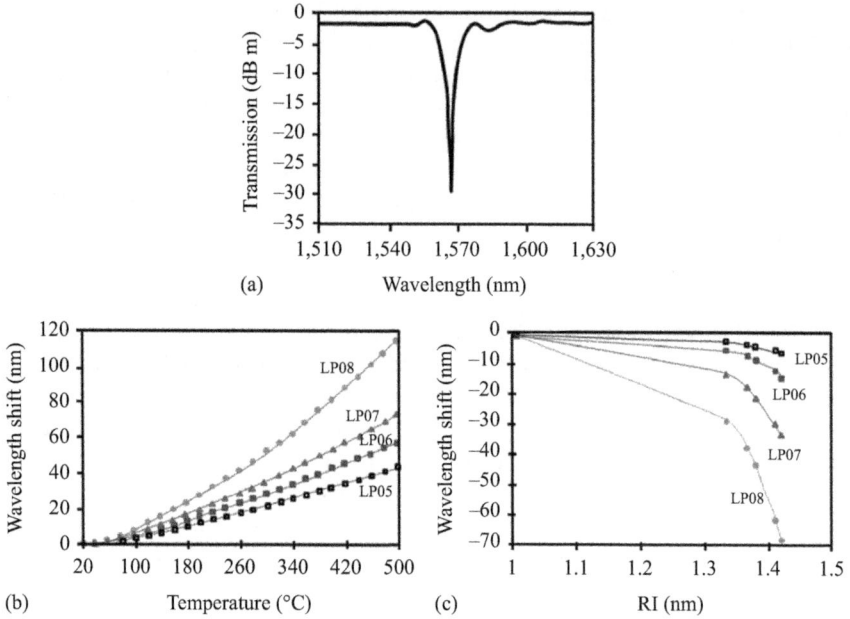

Figure 3.2   Grating resonant wavelength shifted with respect to the temperature: (a) typical example of transmission spectrum, (b) wavelength shift of various FBG sensors as a function of temperature and (c) wavelength shift of various FBG sensors as function of RI

The higher the cladding mode is, the higher the sensitivity of the grating towards RI changes. In addition, the FBG responded to the RI changes quite non-linearly.

As mentioned in Chapter 2, the Bragg grating resonance, which is the centre wavelength of reflected optical signal from a Bragg grating, depends on the effective RI of the fibre core and the periodicity of the grating, will be influenced by variation in temperature.

To obtain the temperature dependence of the grating, we discern the grating equation with temperature and obtain the following equation:

$$\frac{\partial \lambda_r}{\partial T} = 2\left(\frac{\partial n_{co}^{eff}}{\partial T} \times \Lambda + n_{co}^{eff} \times \frac{\partial \Lambda}{\partial T}\right) \tag{3.1}$$

where λ_r is the Bragg grating resonance wavelength, T is the temperature, n_{co}^{eff} is the effective RI of the guided core mode and Λ is the Bragg period.

By further simplification, (3.1) becomes (3.2):

$$\frac{\partial \lambda_r}{\partial T} = 2\left(\frac{\partial n_{co}^{eff}}{\partial T} \times \frac{\lambda_r}{2n_{co}^{eff}} + \frac{\lambda_r}{2\Lambda} \times \frac{\partial \Lambda}{\partial T}\right) = \lambda_r \left(\frac{\partial n_{co}^{eff}}{n_{co}^{eff}\partial T} + \frac{\partial \Lambda}{\Lambda\partial T}\right) \tag{3.2}$$

where '$\partial\Lambda/\partial T$' is the thermal expansion coefficient for the fibre (0.55×10^{-6} for the silica core fibre [9]), and $\partial n_{co}^{eff}/n_{co}^{eff}\partial T$ is the thermo-optical coefficient (8.6×10^{-6} for the silica core fibre [9]).

Equation (3.2) anticipates that the thermally induced variation in the Bragg wavelength is due to the variation of RI of the fibre core and the variation of the grating period. The RI variation is the most dominant effect.

In accordance with (3.2), the theoretical temperature sensitivity at the 1,560 nm Bragg grating wavelength is estimated to be 0.014 nm/°C.

3.3   Distributed thermal sensing

Distributed fibre sensors make it possible to measure, e.g., the temperature, stress or strain along a line that may be several kilometres long, using only a single fibre as both the sensing and information-carrying medium. The physical quantity can be measured at virtually every point in the fibre.

Distributed fibre sensors exploit the precept of localisation, as for the radar technology. In fact, the light detection and ranging (LIDAR) systems can treat the obtained signal in the time or frequency domains. The two common techniques are the optical time-domain reflectometry (OTDR) and incoherent optical frequency-domain reflectometry (IOFDR). Both have advantages and drawbacks in terms of resolution, measurement time, cost and complexity.

Note that the term resolution is defined here as the spacing between distinguishable neighbouring measurement points. In OTDR, the location of the scattering section is directly resolved by the time difference from the light emission to the light detection. However, OTDR approach necessitates a huge peak-power pulsed laser and fast electronics to ensure a good resolution and long-range detection. In IOFDR, a laser diode with a narrow line-width acts in a virtually continuous wave regime. The laser is modulated from direct current to frequencies in MHz region, and for each modulation frequency, a stationary-modulated signal is obtained in the detector. Furthermore, by using IOFDR combined with lock-in approaches, the signal-to-noise ratio (S/N ratio) can be ameliorated considerably compared to OTDR [10], but with the cost of additional complications as well, especially in terms of employed electronics. In addition to the amplitudes, the signal phases require to be recorded with minimal errors, posing specific request on electronics. Besides, a rather elaborated signal processing is needed for the data acquisition.

A number of technologies for fibre optic-distributed temperature measurements exist. For example, LIDAR systems based on OTDR Raman scattering with a range of 10–30 km and a resolution of 4–20 m were reported [11,12]. Figure 3.3 shows a schematic illustration of a distributed FBG-sensing network. Simplicity of design, the availability of short-pulsed high-power lasers and photon-counting techniques make this scheme attractive. Distributed sensors based on Brillouin scattering have also been proposed and tested [13]. In this scheme, it is the frequency shift of the scattered light that is temperature dependent. Especially, the possibility to operate in low loss region of the silica fibre and possible use of erbium-doped fibre amplifier

(EDFA) make this scheme favourable in long-range systems. Ranges up to 85 km and resolutions of 20 m have been achieved. Quasi-distributed combined temperature and strain sensors employing arrays of FBG are already commonly used for some applications [14]. In this scenario, the reflected or transmitted wavelength spectrum depends on the temperature and strain. Figure 3.4(a) and (b) displays cases of FBG Bragg wavelength behaviours as a function increasing and decreasing

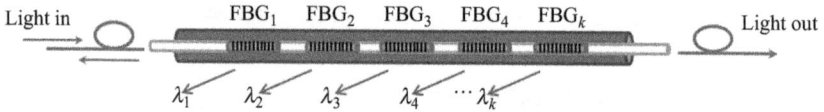

Figure 3.3   *Illustrative description of distributed-sensing network. λ_i corresponds to the reflection wavelength associated to the FBG sensor number i*

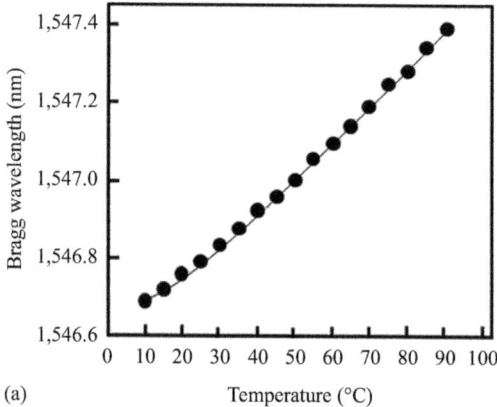

(a)

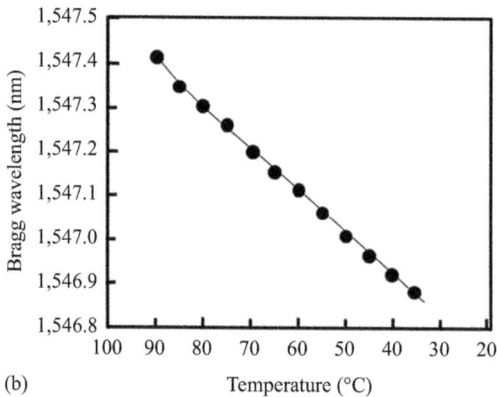

(b)

Figure 3.4   *Bragg wavelength of the FBG as a function of (a) increasing and (b) decreasing temperature*

temperature, respectively. Methods to exploit Rayleigh scattering for distributed temperature sensing have also been found. By analysis of the reflection spectrum of the Rayleigh scattering using coherent optical frequency domain reflectometry, it is possible to detect temperature changes [15,16]. Finally, there are fibres coated with temperature-sensitive coatings, which change transmission characteristics when exposed to temperature changes [17]. All mentioned techniques have, obviously, advantages and limitations compared to one another.

Conventional fibre LPFG have a limited operational temperature of about 150°C because of the temperature-dependent dopant diffusion-induced grating fading. Such fibre gratings break down in more elevated temperature environments. Conventional LPFG exhibit poor thermal drifting whenever temperature >150°C and the grating can be totally extinct when temperature >5,000°C. For operation as a harsh environmental or analytical instrument, it requires the fibre LPFG's RI-modulation amplitude to have high thermal stability. Since the fibre material is normally a floppy structure with a random amorphous network of silicon dioxide with different defects (see the fabrication processes in Chapter 5), temperature change could modify the material's microstructure and morphology. After inscribing a grating in such a fibre material and annealing it at an elevated temperature, the grating structure that is associated with the material microstructure and effective RI in the fibre core can be thermally stabilised.

3.4 Fibre optic sensors doped with rare earth for temperature sensing application

When FOS are doped with rare earth, when they are excited with particular photons energy, they experience the spontaneous stimulated emission of radiation. An in-depth examination of this process was operated for the development of advanced EDFA for optical communication systems having an extended transmission distance [18,19]. The exploration of non-linear phenomena in laser FOS has led the birth of novel optical fibre lasers based on up-conversion process, and a new generation of temperature sensors [20–24].

Commonly, the measurement of temperature based on radiative processes could be of certain advantages, for e.g. there is neither any need of physical/material contact nor the requirement of the equilibrium of the temperature between objects having different thermal masses quantities. Regularly, the temperature is measured indirectly by FOS located at a given distance from the object to be sensed. FOS have already demonstrated their efficiency and accuracy, which is due, among others, to their faculty to transmit light very efficiently and their flexibility permitting them acceding to complicated location and small remote volumes.

Many FOS based on fluorescence processes have been established for temperature sensing, which generally stand on two parameters, namely the fluorescence lifetime and the fluorescence intensity ratio (FIR). The fluorescence-based approaches commonly employ rare earth-doped optical fibres as the sensing materials. The generation of the fluorescence signal is first achieved and then introduced in the fibre by using the continued or pulsed light, with variety of wavelengths. Then, a photodetector is placed to measure the intensity change of the signal of fluorescence with respect to the temperature.

The intensity of the fluorescence signal is generated from a set of doped ions (in the host material) having two energy levels that are nearly spaced. This intensity is dependent on various parameters, including the nature of the host material; the concentration of doping; the energy level of ions; and the selected light excitation method. The separation between the energy levels that are closely spaced may be of the order of a few kT [where 'kT' is about 200 cm^{-1} at room temperature (RT)]. Rare earth-doped materials are among a few numbers of materials that satisfy such energy difference and are hence considered for the thermal coupling, and could potentially be employed with the FIR method for temperature sensing.

An exhaustive review on the rare earth-doped FOS based on the FIR approach can be found in [25–31].

When the FIR approach is used, the thermally coupled levels method presents number of advantages over the two non-coupled levels methods, including

- The fact that the theory of the relative variation in the intensity of the fluorescence – which is originating from the thermally coupled ions levels – is rationally well understood makes the accurate prediction of their behaviour possible.
- As the set of energy levels that are individually thermally coupled is proportionate to the total population of the level itself, any variation in the latter occurring from the variation in the excitation intensity or power will affect the individual levels in the same way; therefore, we can minimise the dependence of the measurement technique on the excitation parameters, reducing thereby the measurements errors.
- For close energy levels (i.e. neighbouring spaced), the wavelengths related to the fluorescence will be as well close. This will consequently reduce the wavelength-dependent effects occurring from the fibre bends.

In these FOS systems, the degree at which the FIR varies with respect to the temperature variation is called sensitivity '$S(R)$', and is given by

$$S(R) = \frac{1}{R}\frac{\partial R}{\partial T} = \frac{\Delta E}{kT^2} \tag{3.3}$$

where R is FIR, T is the temperature, $\partial R/\partial T$ is the rate of change of the FIR with temperature, ΔE is energy difference of a given pair of energy levels and k is the Boltzmann constant.

From (3.3), we can see that when the energy difference of a given pair of energy levels ΔE is larger, the FIR sensitivity increases. However, it is noteworthy to indicate that this amount of energy difference is restricted and limited by the development and the materialisation of thermalisation phenomenon. When this energy difference is more and more larger, the population of levels and then the fluorescence intensity decrease, thereby impacting negatively the accuracy of measuring very low light levels.

Moreover, additional factors are limiting the utility of a material as a sensor, including its availability, cost, the temperature range at which the material aimed to be employed and the fluorescence output. The rare earth ions triply ionised were found to meet all these requirements.

When integrating FOS for thermally sensing, not only the energy levels should be thermally coupled but also the other specifications related to the host matrix materials need to be met. For example, in the case of silica glass, the energy levels have to obey to the following conditions:

- The pair of energy levels must be thermally coupled where the separation level should be below 2,000 cm^{-1}; otherwise, for the temperature interval under study, the population of the top energy level will be very small.
- To avoid overlapping between two adjacent fluorescence wavelengths, the minimum separation should be of 200 cm^{-1} and above.
- The domination of the radiative transitions over the non-radiative ones is a pre-request to generate a good intensity of the fluorescence signal from the pair of upper energy levels. The degree of the non-radiative transition will diminish as a function of increasing the energy gap. Consequently, it is more relevant that the two thermalising levels stand above the next lowest energy level at least at 3,000 cm^{-1}.
- The detectors are generally employed in sensors systems, such as silica photodiodes. In these detectors, the energy levels may show fluorescence (i.e. radiative transitions) with energies ranging from 6,000 to 25,000 cm^{-1}. This corresponds to wavelengths ranging from 1.66 to 0.40 µm, respectively.
- For practical point of view, commercial light-emitting diodes (LED) or laser diodes are better to be used as signal of fluorescence excitation source.

A survey of the relevant literature shows that only a few rare earth ions having a pair of energy levels that meet all the requirements above are available.

Consequently, the rare earth ions that are potentially used as sensing materials for temperature measurements are neodymium (Nd^{3+}), praseodymium (Pr^{3+}), europium (Eu^{3+}), samarium (Sm^{3+}), erbium (Er^{3+}), holmium (Ho^{3+}) and ytterbium (Yb^{3+}). All of them have the ability to be doped into a large range of variety of hosts including glass and/or crystal. The rare earth ions energy levels and their respective fluorescence transitions are detailed in the relevant literature for a variety of host materials. Table 3.1 summarises the performance of fibres based on rare earth-doped materials and employed as temperature sensors using the FIR approach [18,19].

For sensing the temperature, various experimental preparation processes are operating in the FIR method. The fundamental elements employed in this technique are detailed in the following section.

For these rare earth ions, to study their photo-thermal characteristics in various host matrices, the specimen may be irradiated by a laser or pig-tailed diode pump sources that generate the fluorescence from a nearly pair of energy levels as detailed before. The specimen could be then either cooled down or heated up, and their temperature could be identified separately via a thermocouple put in close proximity to the specimen. Subsequently, to record the generated fluorescence

Table 3.1   Performance characteristics of various rare earth-doped-based fibres

Sensing material	Energy gap (cm^{-1})	λ pump (nm)	Temperature range (°C)	Sensitivity at 20°C
Er^{3+}: silica fibre	800	800	23–600	1.3%/°C
Er^{3+}: silica fibre	800	800	25–600	0.8%/°C
Er^{3+} and Er/Yb: chalcogenide glasses	800	1,540 and 1,064	20–220	1.02%/°C, 0.52%/°C at 220°C
Nd^{3+}: silica fibre	1,000	802	−50 to 500	1.68%/°C
Pr^{3+}: silica fibre	580	488	22–250	0.39%/°C
Pr^{3+}: aluminosilicate fibre	580	488	−185 to 257	0.14%/°C
Pr^{3+}: ZBLAN glass	580	450 (LED)	−45 to 255	0.48%/°C
Yb^{3+}: silica fibre	Stark sublevels	810	20–600	0.95%/°C
Er^{3+}: doped fibre	Stark sublevels	1,480	−50 to 90	0.7%/°C
Sm^{3+}: silica fibre	1,000	476.5	22–475	1.85%/°C
Eu^{3+}: silica fibre	1,750	465	−172 to 400	0.178%/°C
Dy^{3+}: silica fibre	1,000	477	21.5–250	1.05%/°C
Er^{3+}: silica fibre	800	975	20–200	3.5%/°C

spectrum and to estimate the calculated intensity ratio with respect to the temperature of the samples, many techniques could be used, including an optical spectral analyser, a photodetector and a band-pass filter.

Practically, compact FOS with a good S/N ratio and high sensitivity are the most suited. To figure out these specifications, a FOS doped with erbium was investigated as a temperature sensor to assess its ability to monitor thermally generated radiation [25,30]. Subsequently, its performance was evaluated experimentally. In the fluorescent sensor, a detection system incorporating two optical channels to select the fluorescence spectral bands which are emitted from levels 2H11/2 and 4S3/2 of the erbium-doped FOS was integrated to depict the temperature-sensing data encoded in the recorded fluorescence spectrum [27,30].

3.5   High-pressure fibre Bragg grating sensing

Upon exposition to high pressures, the FBG sensors experience a negative wavelength shift that is correlative directly to the level of hydrostatic pressure [32]. Figure 3.5 shows a representative example of the Bragg wavelength shift occurring in the grating spectrum undergoing a mechanical strained. The Bragg wavelength λ shifts and becomes λ'. This wavelength shift is rather small and is habitual in the range of −3 Pm/MPa. However, it was established that this 'λ' shift varies linearly with pressures up to 70 MPa. On the other hand, it was proven that when FBG are coated with thermosets polymers like polyurethane or elastomer, their sensitivity to pressure is highly improved up to 20 times [33]. More recently, an FBG adhered with carbon fibre laminated composite structure showed 1,000 times higher sensitivity as demonstrated for hydrostatic pressures up to 70 MPa [34]. The determination of these two cases was performed at RT.

In the case of prerequisite of measuring simultaneously high pressures and high temperatures that are above few hundreds of degree Celsius, the sensing of multi-parameter pressure/temperature sensor employing grating structures is not valid. By using the fibre geometry depicted in Figure 3.6, this side-hole fibre structure allows concurring measurements of high pressure and temperature at the same time [8,35]. The birefringent configuration of the fibre is ending in two kind of polarisation, which are both dependent on the Bragg grating resonances frequency. Hence, by increasing the pressure inside the side holes, the separation of the polarisation-dependent resonances is reduced, and the variation of any temperature will be detected by the same wavelength shifts of the two resonances.

A FOS able to measure simultaneously temperature and pressure in extreme harsh environments is possible by using this particular configuration that produce a femtosecond infrared (IR) laser-induced FBG that is thermally stable and regenerated

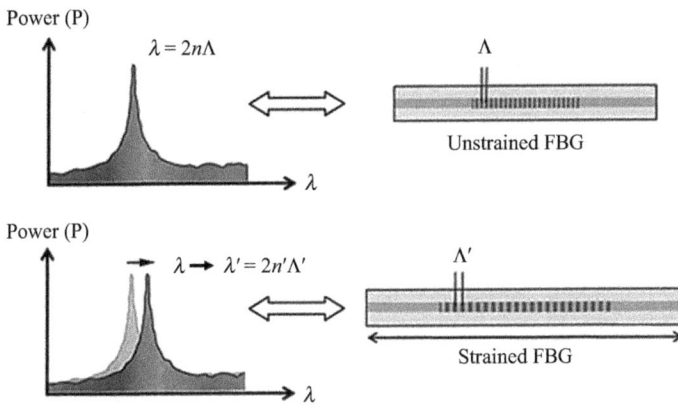

Figure 3.5   Typical example of occurred shift in the grating spectrum power (P) undergoing a mechanical strained. Λ and Λ' are the Bragg periods before and after the mechanical strain, respectively, and n is an integer

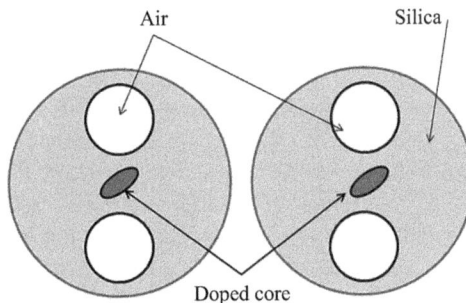

Figure 3.6   Cross section of the side-hole fibre with tilted elliptical core

gratings engraved in microstructured side-hole fibre [36,37]. These FOS were already established to operate efficiently under pressure ranging from 15 to 2,400 psi (0.1–16.5 MPa) and temperatures above 800°C. This configuration could be also employed to estimate the RI of materials and even fluids inside the side hole.

Recently, there has been an increasing interest towards the usage of hydrogen in the field of clean energy production, combustion engine vehicles, petrochemical and fuel cell technology. The latter is being seen as the most promising technology application.

Among of the main advantages by using hydrogen is the fact that it does not engender any pollutants during the combustion process and can be originated from environmental friendly sources. Hydrogen is commonly stored inside high pressure tanks.

As a matter of fact, in the nuclear industry, advanced reactors, such as the next generation very high-temperature and supercritical water-cooled reactors, are expected to be potentially highly efficient processes for the large production of hydrogen. However, recommended temperatures and pressures in these processes are in the ranges of 1,000°C and 25 MPa. Consequently, to detect hydrogen in both ambient and harsh conditions, using various hydrogen sensors is a mandatory condition.

Among these requirements, sensors are requested to be small, cost effective, robust and particularly reliable over a large range of temperature, in addition to the long-term stability, reproducibility and operability in environment with potentially explosive conditions.

A particular type of FOS, namely FBG sensor coated with thin palladium layer was studied in-depth for hydrogen detection for low-volume concentrations (in the range of 1–4 vol.%) [38].

The physics behind is to use either the volume variation of the Pd layer that generate strain resulting thereby in spectral shift of the Bragg wavelength that FGB could monitor, or the variation of the Pd RI [38,39]. The latter is true when side polished fibres are vanishingly coupled into the Pd layer. To improve the responsiveness of the sensor, the FBG can be heated up.

It is noteworthy to indicate that the palladium–hydrogen system dependents on the hydrogen pressure and temperature. Pd and H could be in two different crystallographic phases, namely α and β, which are separated by a transition phase. It is then imperative to run the sensor in the appropriate range of temperature pressure of one or the other of these two phases [40].

To improve the photosensitivity to UV light of Ge-doped silica waveguides, the hydrogen loading process is often employed [41]. This loading process could be employed as well as a hydrogen sensor. The loading mechanism typically necessities to put the optical fibres inside an atmosphere of H_2 for 1–2 weeks at RT and ~2,000 psi, or for 1–2 days at higher temperatures of 80°C–100°C.

From chemical point of view, the hydrogen will not react with the fibre material but is rather in 'solution' with the silica matrix. Once the optical fibre is removed from the high pressure H_2 environment, the hydrogen within the fibre starts to outgas instantly [41]. The role of loaded hydrogen in the fibre is limited to triggering a change in the RI (of the fibre core) which tends to increase and is subject to the temperature and the partial pressure of the H_2. Typically, the H_2 loading leads to a red-shift (i.e. longer wavelength) of the Bragg resonance of 1 nm during the time fibre is under pressure. During the out gassing process, the resonance of the Bragg grating shifts and returns back to its initial wavelength value. The duration of this phenomenon (i.e. red–blue-shift within and without the presence of H_2) is somehow too long for the sensing operation. However, by using appropriate optical fibres, for e.g. photonic crystal-based fibres or side-hole fibres, this reaction time could be significantly reduced. These two kinds of fibres have the ability to bring the hydrogen in proximity of the fibre core, i.e. in a more direct contact with the ambient H_2 gas.

For the rates of out-diffusion of high-pressure hydrogen gas from photonic crystal fibre cores using Bragg gratings [42].

3.6   High reliability of the fibre Bragg gratings for the high strain measurements

Measuring large strain levels for the monitoring of the structural integrity by using an array of FBG sensor is highly desirable and useful. The standard grating writing through UV irradiation process requires first removing the fibre-protective polymer coating (called jacket) because it is highly absorbing UV light. Then, this jacket should be replaced after the FBG inscription. This operation of coating removing/ replacing is time-consuming and may affect the mechanical integrity of the fibre which in turn reduces the sensitivity to the strain measurement.

It has been shown that by using the a phase mask technique combined with femtosecond IR laser, the gratings were successfully inscribed without removing the acrylate coating of standard Corning® SMF-28® fibre [43]. On the other hand, the photosensitivity of the Ge-doped fibre to femtosecond IR laser was considerably improved by H_2 loading, and a high numerical aperture (NA) was as well achieved [44,45]. A modulation of the RI of 1.4×10^{-3} was successfully achieved in these high-NA fibres, with strengths kept at 75%–85% of the original fibre value.

However, for applications requiring accurate strain sensing using bend insensitive fibres, it is preferable to have an FBG coated with thin polyimide, as it more sensitive to transfer the mechanical deformation to the fibre than the thick acrylate coatings. Employing this same process in polyimide-coated bend insensitive high-NA fibres, an index modulation of 1×10^{-4} has been successfully generated, while the strength remained at ~50% of the pristine fibre value [46].

3.7   Fibre Bragg gratings exposed to high-dose radiation

Recently, the capability of FOS to operate in high-temperature and high-radiation environments conditions has been established, and particularly under high-gamma radiation doses [47–50]. Many relevant works have also proven that hydrogen-loaded Ge-doped FBG sensors are particularly able to undergo high-gamma radiation doses [51,52]. Globally, unlike physical radiation sensors based on electrical signal of electronic dosimeters, the main advantage of the FBG sensor is that the information of the radiation dose is transmitted using optical signals and is consequently immune against the electrical and EMI. In addition, the electrical dosimetry employs a high-voltage power supply and should have good electrical insulation. However, degradation occurs over time to the insulation upon irradiation and may affect considerably both the stability and the good functioning of the dosimeter. Moreover, one has to take into account the electronic noise generated by radiation, which affects the accuracy of the signal-transmitting information [49,52].

Silicon dioxide (SiO_2) is currently the most used material for the optical fibre fabrication. It is also known as pure glass [53]. Other materials are also in use, including plastic-based fibre and fluoride-based glass fibre; however, they are less employed because of their multiple disadvantages comparatively to the silicon dioxide ones. For example, the plastic optical fibres show high signal attenuation and are consequently not suitable as FOS especially in radiation environments [53].

Upon irradiation, especially under high doses, centres of defects could happen in the silica material of optical fibres. These defects centres – also called colour centres – are the main factor increasing the attenuation loss [54]. In the same way, under irradiation, the RI of the optical fibre also varies owing to the physical damage occurring in the structure of the fibre matrix leading thereby to increasing the fibre-absorption loss [55]. These physical defects will induce new energy levels within the band-gap increase, hence the absorption of the optical transmitted signal. This phenomenon is known as radiation-induced attenuation (RIA) [56,57].

However, doping the fibre with germanium, boron, erbium, phosphorus and fluorine improves the light propagation and reduce considerably the attenuation [57,58]. The most employed doped fibre nowadays is based on the germanosilicate glass [57–59].

Both doped (Ge-) and undoped silicate glass manifest an attenuation rate of about 0.20 dB/km only, measured at a wavelength of 1,550 nm [58–60]. Ge is also used as a dopant to enhance the RI between the fibre core and cladding, ensuring thus improved light guiding characteristics [47].

Germanosilicate glass fibres are also known to be highly photosensitive. This photosensitive property could be even enhanced further by loading hydrogen and/or by boron co-doping [61]. When using both Ge doping and H_2 loading, the radiation sensitivity of the fibre is also substantially increased. It is noteworthy that the RI of the core should be higher than that of the cladding to obtain a light propagation inside the fibre. As a matter of fact, Figure 3.7 shows the propagation of light wave in the optical fibre showing high and low RI.

As for the temperature gradient and mechanical strain, the gamma irradiation exposition of the FBG led to a clear shift of the Bragg wavelength (Figure 3.8). Results show as well that when FBG are exposed to gamma radiation with a total dose of 100 kGy, their sensitivity increases from 820 to 1,516 nm. FBG sensors with wavelength above 1,550 nm are more suitable to sense high dose of radiation, as no saturation was observed up to 100 kGy [62].

Thus, the examination has revealed that temperature and the radiation-induced Bragg wavelength shift could be estimated with high exactitude. It has been

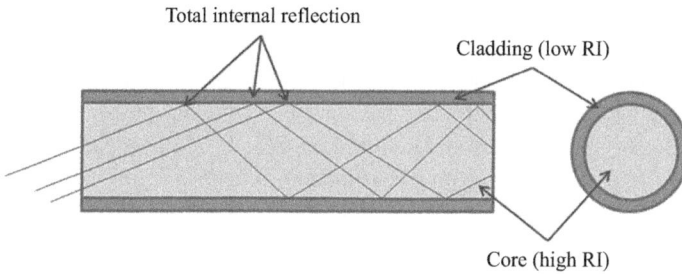

Figure 3.7   Propagation of light in the optical fibre showing high and low RI

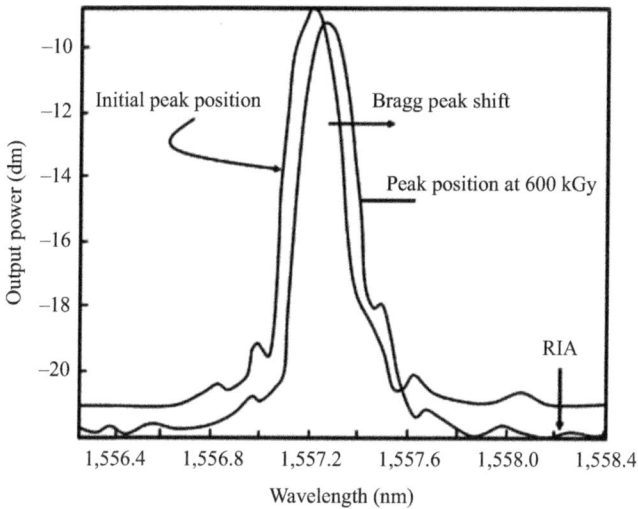

Figure 3.8   Radiation-induced Bragg peak shift, with baseline lowered due to RIA.
© 2002 IEEE. Reproduced, with the permission, from [63]

demonstrated that the effect of hydrogen load in the FOS is crucial, where the gamma radiation sensitivities formed by a 330 nm UV light in Ge-doped fibre that are hydrogen loaded are much higher than those without hydrogen loading. In addition, when the FBG is formed in Ge-doped fibre loaded with H_2, the Bragg wavelength shift peak saturated at a higher level of radiation, in comparison with the same fibres with no hydrogen loading [50].

In sum, the FBG sensitivity to the radiation is subject to the level of photosensitisation and the chemical composition used for their fabrication. In fact, FBG sensors which have been formed of a fibre which is naturally photosensitive are very tolerant to radiation and show a small Bragg wavelength peak shifting. When it is a matter of gamma radiation, this shift occurs typically towards the longer wavelengths [64]. Various fabrication methods lead to various Ge content in the fibres.

When the fibres are H_2 loaded, the Ge–silica fibres undergo a Bragg wavelength shift of about 160 pm upon an exposure to 100 kGy dose of gamma radiation, and of about 50 pm only when they are not Ge doped [58]. Fibres with about 10–21 wt% Ge doping are shown to be the best suited as sensors for radiation dosimetry [59].

Gusarov [65] wrote an extensive study on the effects of long-term exposure to radiation in a nuclear reactor on FBG sensors, where the gratings networks are written in a typical photosensitive and standard optical fibres. This elegant investigation was conducted over a period of 8 years. During that time, the nuclear reactor was operational for a total of 4,690 h. For unloaded fibres (i.e. without hydrogen) and after the 8 years of gamma exposure, it was found that the amplitude and the gratings spectra remained unchanged. The fibres with hydrogen loading revealed only minor variations to the amplitude and the grating spectra [65]. This result implied that the Bragg gratings could tolerate long-term exposure without serious degradation of the photoreflectivity and with only an insignificant shift of the Bragg gratings [65]. These findings are crucial as the majority of previous experiments operated to assess the FBG behaviour in harsh conditions were conducted in relatively brief time scale (often a matter of days or weeks). These investigations have been performed similar to 8-year study. A 50-month experiment was also tackled to demonstrate the long-term effects on FBG-based temperature sensors in a low-flux nuclear reactor [66], and the obtained results revealed that the sensors continually operate efficiently and can tolerate long-term exposure to moderate nuclear radiation. The temperature sensitivity capability was insensitive to the long-term irradiation exposure, allowing then to perform a temperature measurement variation with a precision of less than 3°C, which is an accepted standard for nuclear application measurement [66].

3.8 Conclusions

In this chapter, we reviewed some of the most contemporary advancements in the application of FBG sensors in extreme environment conditions. Although the dominant technologies used today to perform temperature measurements are based on electrical sensing, optical sensors may present a tremendous offer as a relevant alternative for challenging applications in harsh environments. As a matter of fact, distributed systems, space frontiers and long-term implementation are typical domains where the unprecedented FBG characteristics make a clear difference by achieving a highly effective solution as compared to conventional electrical sensors.

References

[1]   W. Liang, Y. Huang, Y. Xu, R.K. Lee and A. Yariv, *Applied Physics Letters*, 2005, **86**, 151122.

[2]   A. Othonos and K. Kalli in *Fibre Bragg Grating: Fundamentals and Applications in Telecommunication and Sensing*, Artech House, Boston, MA, USA, 1999.

[3]   K. Hill and G. Meltz, *Journal of Lightwave Technology*, 1997, **15**, 1263.

[4]   T. Erdogan, V. Mizrahi, P. Lemaire and D. Monroe, *Journal of Applied Physics*, 1994, **76**, 73.

[5]   S. Trpkovski, D.J. Kitcher, G.W. Baxter, S.F. Collins and S.A. Wade, *Optics Letters*, 2005, **30**, 607.

[6]   B. Zhang and M. Kahrizi, *IEEE Sensors Journal*, 2007, **7**, 586.

[7]   J. Canning, M. Stevenson, K. Cook, *et al.*, *Proceedings of SPIE*, 2009, **7503**, doi:10.1117/12.834470.

[8]   S.J. Mihailov, *Sensors*, 2012, **12**, 2, 1898.

[9]   W. Zhao and R.O. Claus, *Smart Materials and Structures*, 2000, **9**, 212.

[10] M.A. Farahani and T. Gogolla, *Journal of Lightwave Technology*, 1999, **17**, 1379.

[11] P. Rajeev, J. Kodikara, W.K. Chiu and T. Kuen, *Key Engineering Materials*, 2013, **558**, 424.

[12] Z. Zaixuan, L. Honglin, G. Ning, *et al.*, *Proceedings of SPIE*, 2003, **4893**, 78.

[13] Y.T. Cho, M. Alahbabi, G. Brambilla and T.P. Newson, *Proceeding of the Conference on Lasers and Electro-Optics*, 2004, **1**, 2.

[14] A.D. Kersey, M.A. Davis, H.J. Patrick, M. LeBlanc and K.P. Koo, *Journal of Lightwave Technology*, 1997, **15**, 1442.

[15] L. Palmieri and L. Schenato, *The Open Optics Journal*, 2013, **7**, 104.

[16] Y. Koyomada, Y. Eda, S. Hirose, S. Nakamura and K. Hogari, *IEICE Transactions on Communications*, 2006, **E-89B**, 1722.

[17] S.M. Chandani and N.A.F. Jaeger, *IEEE Photonics Technology Letters*, 2005, **17**, 2706.

[18] E. Desurvire in *Erbium Doped Fibre Amplifiers: Principles and Applications*, 1st Edition, John Wiley & Sons Inc. Publisher, New York, NY, USA, 1994.

[19] M.J.F. Digonnet in *Rare Earth Doped Fibre Lasers and Amplifiers*, 2nd Edition, Marcel Dekker Inc. Publisher, New York, NY, USA, 2001.

[20] E.B. Mejia, A.A. Senin, J.M. Talmadge and J.G. Eden, *IEEE Photonics Technology Letters*, 2002, **14**, 11, 1500.

[21] D.V. Talavera and E.B. Mejia, *Journal of Applied Physics*, 2005, **97**, 5, 053102.

[22] H. Berthou and C.K. Jorgensen, *Optics Letters*, 1990, **15**, 19, 1100.

[23] M.C. Farries, M.E. Fernmann, R.I. Laming, S.B. Poole, D.N. Payne and A.P. Leach, *Electronics Letters*, 1986, **22**, 8, 418.

[24] P.A. Krug, M.G. Sceats, G.R. Atkins, S.C. Guy and S.B. Poole, *Optics Letters*, 1991, **16**, 24, 1976.

[25] J. Castrellon and G. Paez, *Proceedings of SPIE*, 1999, **3759**, 410.

[26] J. Castrellon-Uribe and G. Garcia-Torales, *Fibre and Integrated Optics*, 2010, **29**, 4, 272.

[27] J. Castrellon-Uribe, *Optics and Lasers in Engineering*, 2005, **43**, 6, 633.

[28] J. Castrellon-Uribe, C. Cuevas-Arteaga and A. Trujillo-Estrada, *Optics and Lasers in Engineering*, 2008, **46**, 6, 469.

[29]   J. Castrellon-Uribe, M.E. Nicho and G. Reyes-Merino, *Sensors & Actuators: B: Chemical*, 2009, **141**, 1, 40.

[30]   J. Castrellon-Uribe, G. Paez and M. Strojnik, *Optical Engineering*, 2002, **41**, 6, 1255.

[31]   J. Castrellon-Uribe, G. Paez and M. Strojnik, *Infrared Physics & Technology*, 2002, **43**, 3–5, 219.

[32]   M.G. Xu, L. Reekie, Y.T. Chow and J.P. Dakin, *Electronic Letters*, 1993, **29**, 398.

[33]   D.J. Hill and G.A. Cranch, *Electronics Letters*, 1999, **35**, 1268.

[34]   Z. Wei, D. Song, Q. Zhao and H.-L. Cui, *IEEE Sensors Journal*, 2008, **8**, 1615.

[35]   S.J. Mihailov, D. Grobnic, H. Ding, C.W. Smelser and J. Broeng, *IEEE Photonics Technology Letters* 2006, **18**, 1837.

[36]   C.M. Jewart, Q. Wang, J. Canning, D. Grobnic, S.J. Mihailov and K.P. Chen, *Optics Letters*, 2010, **35**, 1443.

[37]   T. Chen, R. Chen, C. Jewart, *et al.*, *Optics Letters*, 2011, **36**, 3542.

[38]   R.R.J. Maier, B.J.S. Jones, J.S. Barton, *et al.*, *Journal of Optics A: Pure and Applied Optics*, 2007, **9**, 6, S45.

[39]   D. Zalvidea, A. Díez, J.L. Cruz and M.V. Andrés, *Sensors and Actuators B: Chemicals*, 2006, **114**, 268.

[40]   A. Trouillet, E. Marin and C. Veillas, *Measurement Science and Technology*, 2006, **17**, 1124.

[41]   B. Malo, J. Albert, K.O. Hill, F. Bilodeau and D.C. Johnson, *Electronics Letters*, 1994, **30**, 442.

[42]   T. Geernaert, M. Becker, P. Mergo, *et al.*, *Journal of Lightwave Technology*, 2010, **28**, 1459.

[43]   S.J. Mihailov, D. Grobnic and C.W. Smelser, *Electronics Letters*, 2007, **43**, 442.

[44]   D. Grobnic, S.J. Mihailov, C.W. Smelser and R.T. Ramos, *IEEE Photonics Technology Letters*, 2008, **20**, 973.

[45]   C.W. Smelser, S.J. Mihailov and D. Grobnic, *Optics Letters*, 2004, **29**, 2127.

[46]   S.J. Mihailov, D. Grobnic, R.B. Walker, *et al.*, *Optics Communications*, 2008, **281**, 5344.

[47]   A. Stanciu and M. Stanciu, *Electrotehnica Electronica Automatica*, 2005, **53**, 4, 15.

[48]   A. Gusarov, B. Brichard and D.N. Nikogosyan, *IEEE Transactions on Nuclear Science*, 2010, **57**, 4, 2024.

[49]   T. Shikama, K. Toh, S. Nagata and B. Tsuchiya, *Proceedings of SPIE*, 2005, **5855**, 507.

[50]   A.I. Gusarov, F. Berghmans, A. Fernandez Fernandez, *et al.*, *IEEE Transactions on Nuclear Science*, 2000, **47**, 3, 688.

[51]   H. Henschel, D. Grobnic, S.K. Hoeffgen, J. Kuhnhenn, S.J. Mihailov and U. Weinand, *IEEE Transactions on Nuclear Science*, 2011, **58**, 4, 2103.

[52]   D. Grobnic, H. Henschel, S.K. Hoegggen, J. Kuhnhenn, S.J. Mihailov and U. Weinand, *Proceedings of SPIE*, 2009, **7316**, 73160C.

[53] S. O'Keefe, C. Fitzpatrick, E. Lewis and A.I. Al-Shamma, *Sensor Review*, 2008, **28**, 2, 136.

[54] S. Agnello, R. Boscaino, M. Cannas and F.M. Gelardi, *Journal of Non-Crystalline Solids*, 1998, **232**, 328.

[55] A. Morana in *Gamma-rays and Neutrons effects on Optical Fibers and Bragg Gratings for Temperature Sensors, Optics/Photonic*, Jean Monnet University, Saint-Étienne, University of Palermo, Palermo, Italy, 2013.

[56] S. Girard, Y. Ouerdane, G. Origlio, *et al.*, *IEEE Transactions on Nuclear Science*, 2008, **55**, 6, 3473.

[57] S. Girard, N. Richard, Y. Ouerdane, *et al.*, *IEEE Transactions on Nuclear Science*, 2008, **55**, 6, 3508.

[58] S. Lin, N. Song, J. Jin, X. Wang and G. Yang, *IEEE Transactions on Nuclear Science*, 2011, **58**, 4, 2111.

[59] H. Henschel, S.K. Hoeffgen, K. Krebber, J. Kuhnhenn and U. Weinand, *IEEE Transactions on Nuclear Science*, 2008, **55**, 4, 2235.

[60] E.M. Dianov and V.M. Mashinsky, *Journal of Lightwave Technology*, 2005, **23**, 11, 3500.

[61] A. Othonos, *Review of Scientific Instruments*, 1997, **68**, 12, 4309.

[62] G.P. Agrawal in *Fibre-Optic Communication Systems*, 4th Edition, John Wiley & Sons. Inc. Publisher, New York, NY, USA, 1997.

[63] A. Fernandez, B. Brichard, F. Berghnans and M. Decreton, *IEEE Transactions on Nuclear Science*, 2002, **49**, 6, 2874.

[64] A. Fernandez Fernandez, A. Gusarov, F. Berghmans, *et al.*, *Proceedings of SPIE*, 2002, **4823**, 205.

[65] A. Gusarov, *IEEE Transactions on Nuclear Science*, 2010, **57**, 4, 2044.

[66] A. Fernandez-Fernandez, A. Gusarov, B. Prichard, *et al.*, *Measurement Science and Technology*, 2004, **15**, 1506.

Chapter 4

Main applications of optical fibres and fibre Bragg grating sensors

Following the enormous development of the fibre Bragg grating(s) (FBG) sensor technology in the recent years, different kinds of fibre optic sensors (FOS) with advanced faculties were designed and matured to sense various parameters, including temperature, radiation, rotation, strain, acceleration, vibration, humidity, refractive index (RI) and so on. FOS have become more and more important and even indispensable for many industries, such as in civil, medical, military and space applications. In the following chapter, we will detail and review the main applications of FBG sensors while putting focus on the communications field.

4.1 Classification of fibre optic sensors

Nowadays, numerous kinds of FOS have been successfully developed and find their way towards the 'real-world' applications. Basically, FOS consist of four major components: optical source, signal waveguide (e.g. optical fibre), sensing head and detector (Figure 4.1). Light propagates along an optical fibre passing through the sensing head, where certain properties of the light wave would be modulated when the sensing head is influenced by the external variations which need to be monitored. The modulated light wave (containing the information) is then received by the detector, and after the demodulation, the external varying parameters which need to be measured could be obtained.

Conventionally, FOS are gathered into two main categories, namely the intrinsic and extrinsic sensors. For extrinsic sensors, measuring the parameter to sense occurs in a region located out of the fibre.

Figure 4.2 shows the case of extrinsic FOS. The optical fibre implies a so-called black box, which links and correlates the information which needs to be monitored onto the optical signal, in echo to media effects. The fibre then guides the optical signal containing the sensed information by the surrounding environment back to detector [1]. The black box usually contains other electrical or optical devices; hence, the extrinsic sensors also refer to as hybrid FOS.

In the case of intrinsic sensors, the physical characteristics of the fibre will undergo a change under environment variation, where the optical signal that carries out the sensing information will be affected proportionally upon external environmental change. The illustration is shown in Figure 4.3.

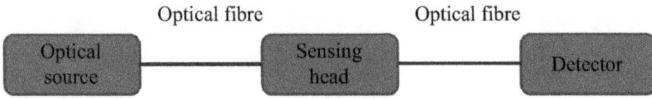

Figure 4.1 *Principle of the FOS configuration, consisting of optical source, optical fibre, sensing head and detector*

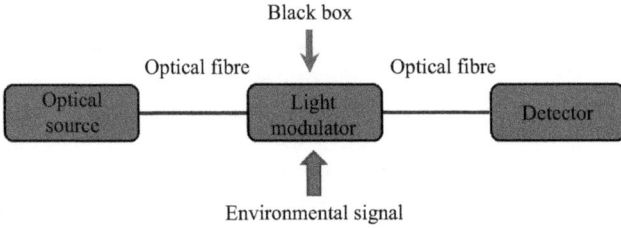

Figure 4.2 *Extrinsic FOS consisting of the optical fibre that leads up to and out of a black box that modulated the optical signal passing through it under environment change*

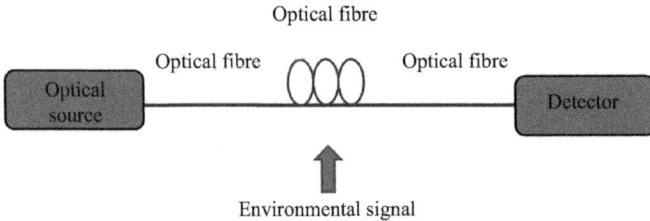

Figure 4.3 *Intrinsic FOS using the propagating of the optical signal through the optical fibre. The light beam is modulated by the various environmental effects*

The intrinsic FOS could also refer to as all-fibre sensors. However, different other classification methods for FOS exist in order to be as convenient as possible to the users (researchers or engineers). Since there are a variety of kinds of FOS, it is hard to make a very comprehensive illustration.

The following subsections only introduce some basic categorisations.

4.1.1 Categories according to fibre Bragg grating application

Based on the parameters the FBG sensor target to measure, or according to different applications, FOS could be categorised as temperature sensors, strain sensors, displacement sensors, current sensors, pressure sensors, vibration sensors, chemical sensors and so on.

This is a straightforward method to categorise fibre sensors. The users could find the FOS easily to meet their requirements. For most of the FOS manufactures, they advertise their sensor products according to applications in order to become 'user friendly'.

4.1.2   Categories according to measurable spatial scope

Based on the spatial scope as such, the FBG sensors could also be categorised as point sensors, quasi-distributed sensors and fully distributed sensors.

4.1.2.1   Point sensors

The small size is one of the main strengths of the FOS. The point sensors could be used to measure the parameters of certain points in the space. As a matter of fact, in medical application, temperature and pressure inside certain tissues are important information which need to be monitored dynamically during the surgery. Owing to their miniature size, the FBG sensors become more and more important in such applications.

In addition to the FBG [2], many other optical fibre components could serve as point sensing head, including optical fibre Fabry–Pérot (FP) interferometers [3] and multimode fibre interferometers [4].

4.1.2.2   Quasi-distributed sensors

Quasi-distributed sensors are able to measure the parameters of many separate points in the space simultaneously, by multiplexing the point sensors to form a sensing mesh. For instance, if people are interested in certain points on some massive structures, or only some positions need to be monitored, by space-division multiplexing FBG and placing the FBG on these positions, all the changes on these points could be monitored simultaneously. Indeed, as the information of the sensed parameter (temperature, radiation, strain and so on) is directly encoded into the optical signal, and more specifically into the light wavelength, this could be successfully achieved. Many quasi-distributed sensing applications with FBG arrays have already been reported, but with different demodulation techniques, such as using filters based on FP [5,6], acousto-optic [7] and FBG [8]. However, for monitoring the health of some massive structures, the locations which need to be monitored are usually not known. This requires the sensors to have the ability for fully distributed sensing, by embedding the sensing fibre into the structure. Although point sensors could realise quasi-distributed sensing, the fact that they are spatially located with limit their capacity to deliver only partial information on the structure health. Hence, there is a clear need for a sensing process that is able to detect faults and evaluate the level of damage of the entire structure of interest. This system should then achieve a fully distributed temperature and strain measurements over large distances.

4.1.2.3   Fully distributed sensors

In this case, to realise this purpose, the fully distributed fibre sensors are often based on Rayleigh [9], Raman [10] and Brillouin scattering [11].

For example, the technique using Brillouin scattering, also called Brillouin optical time domain analysis (BOTDA), is showing very good feature to realise health monitoring of the massive structures and has already been commercialised [11].

4.1.3   Categories according to the modulation process

Along the optical fibre, the light wave propagating is mainly characterised based on four factors, namely the intensity, the phase, the state of polarisation and the wavelength (or frequency). When the media in which the optical fibre is operating undergoes a disturbance on the sensing head, at least one of the four mentioned factors is affected proportionally to the intensity of the environment perturbation. By measuring the variation of the optical signal, the needed information of the surrounding environment change is obtained (i.e. sensing). Therefore, FOS could be categorised according to the modulation process as intensity modulation sensors, phase modulation sensors, polarisation modulation sensors and wavelength modulation sensors.

4.1.4   Categories according to technology

Another straightforward way to categorise FOS is according to different technologies, such as fibre grating sensors [2], fibre FP interferometer sensors [3], high-birefringence fibre-loop mirror sensors [12], polarisation optical time domain reflectometry sensors [13,14], BOTDA sensors [11,15] and so on.

Since there are so many technologies nowadays which could find applications as FOS, it is hard to list them all. However, over the last 40 years, as the optical-fibre-based communication systems have probably impacted our daily life the most, focus is put on the following section on their communication application.

4.2   Advances in optical fibre communication systems

The colossal success met by the optical fibre-based communication technology is undoubtedly due, among others, to their affordable cost.

Historically, the initial phase of research and development (R&D) on optical fibre communication systems commenced in 1975 [16]. The early prototype of optical wave systems was based on gallium arsenide laser and was operating around 0.8 μm. After extensive development and various attempts (especially between 1977 and 1979), these systems became commercially available in 1980 [16]. The first commercial optical fibre communication system operated at a bit rate of 45 Mb/s and granted a repeater spacing of up to 10 km distance.

This repeater spacing distance of 10 km was considered as enormous in comparison with that ensured by coaxial systems, which was limited to 1 km distance. This performance was of tremendous motivation for engineers and system designers as it will decrease considerably the various costs related to the installation and maintenance associated with each repeater.

In the early 1980s, the second generation of optical fibre communication system became commercially available as well, offering a bit rate of about 100 Mb/s.

This limitation of the debit was imputed to the multimode fibres dispersion and was quickly fixed by using single-mode fibres (SMF) [17].

One year after, in 1981, experiment conducted in a laboratory established a data transmission at a bit rate of 2 Gb/s over a distance as long as 44 km. This technological success was achieved by using an SMF [18]. The commercialisation of such a system quickly followed.

In 1987, optical fibre communication system (second generation) running at 1.7 Gb/s and with a repeater spacing over 50 km distance was commercialised.

In 1990, the third generation systems performing at 2.5 Gb/s became accessible, and some of them were even capable of operating at 10 Gb/s [19]. The combination of lasers oscillating in a single longitudinal mode with dispersion-shifted fibres was responsible for such a performance.

A weakness of this third generation systems operating at 1.55 μm is the fact that the optical signal is regenerated periodically based on electronic repeaters that are spaced by 60–70 km.

The fourth optical fibre communication system generation employed optical amplification to even improve the repeater spacing. To increase the bit rate debit, this generation uses the concept of wavelength-division multiplexing (WDM).

In 1996, by using submarine cables, data transmission was successful over a range of 11,300 km and at a debit of 5 Gb/s [20], making thereby transatlantic and transpacific optical fibre communication system commercially available.

The fifth generation of light wave systems was disturbed with continuing the extension of the wavelength range over which a WDM system can operate at the same time.

The commonly established light-wavelength interval, known as the C-band window, is covering the range between 1.53 and 1.57 μm. It was then extended to the L- and S-bands, respectively, i.e. towards the long- and short-wavelength sides. On the other hand, the Raman amplification process could be used for signals in all three C-, L- and S-bands.

In addition, a novel type of fibre was developed, known as the dry fibre, with the aim to diminish the signal losses over the integrated wavelength band lengthening from 1.30 to 1.65 μm [21].

In these dry fibre systems, passive optical network was initially suggested for the formerly conceived fibre to the home (FTTH) network that relies mainly on the use of passive optical splitters, assembled from standard SMF.

Despite the fact that FTTH was not deployed at a large scale until too recently, R&D for the employment of advanced devices for telecommunications applications has still been pursued.

As a matter of fact, in 1990, the optical fibre amplifier was introduced commercially and completely remodelled the technology of the optical fibre transmission. In fact, through amplification, the light signals may transit hundreds of km without any need of regeneration [22].

Two factors control the performance and reliability of a communication system, namely the bandwidth it offers and the level of the received signal-to-noise ratio.

This restriction could be settled in a more formal way through what is currently known as the 'channel capacity' which was introduced by the concept of information theory [23].

The fibre optical network scheme could be notably ameliorated in terms of flexibility and also the capacity of the optical transmission by using the WDM systems [24].

On the other hand, mainly for financial and economic reasons, transmission communication based on the mature technology of SMF optic that offers a high-information capacity will be integrated in future telecommunications networks.

A fascinating aspect for these future optical networks is their capacity to treat the information straightforward in the optical domain, through amplification, correlation, filtering and multiplexing/demultiplexing purposes.

In communications, and in comparison to the information carried out by the electrical signal (the concept of photon–electron–photon conversions), one of the main advantages of the optical signal processing is its ability to travel into a much faster way.

Recently, novel classes of optical networks have been developed [25], including the code-division multiple access networks which are based on the optical signal processing techniques [26]. In addition, the transmission medium is the most used optical fibre technology in telecommunications industry because of its various promising performances beating even those offered by satellite communications [27]. This is mainly due to its incomparable propagation symmetry faculty – in the two directions – that is shown by the optical fibres.

Many advanced industrial projects are dedicated to use the optical fibres transmitting the optical signal which is modulated by the electrical ones. Electrical signals are provided from an atomic clock [28] or from the one which is known as 'coherent optical carrier' induced by classical optics [29]. All over the world, efforts are extensively launched to set up an international fibre optic network.

The complexity of an optical fibre system can vary from a basic case (e.g. local area network) to a very advanced and costly one (e.g. long-distance television/telephone cable trucking). As a matter of fact, the system displayed in Figure 4.4

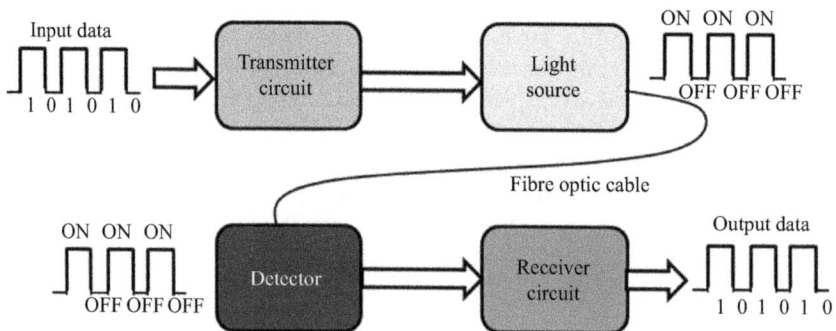

Figure 4.4 Schematic of the basic optical fibre communication system

could be achieved cost effectively by means of light-emitting diode, plastic fibre, a silicon photodetector and basic electronics.

The optical fibres employed for communication can be divided into two main categories, namely single mode and multimode.

Because they have a sharp and clearly defined boundary between the core and the cladding, these fibres show a well-defined RI along the entire core.

Conventionally, a core diameter of 8–9 μm characterises the SMF core, allowing thereby only one mode of light propagation. In contrast, many propagation modes of the optical signal are permitted for the multimode fibre which shows a typical core diameter of 50–62.5 μm and even above. In this latter case, and because of the difference in the travelling distance, the modal dispersion phenomenon could occur for some particular propagation modes that undergo longer time to travel the fibre than others.

To reduce this modal dispersion factor, the multimode-graded index fibre is often used. In this particular fibre, and starting from the centre of the fibre core, the RI is deliberately reduced. Consequently, the light propagation is increasingly refracted starting from the centre of the fibre, which decreases the speed of specific light waves, hence allowing all the optical signals modes to reach the photodetector at almost simultaneously. Figure 4.5 displays a schematic of the two main optical fibre index modes, namely the step index (single and multimodes) and graded index (parabolic profile).

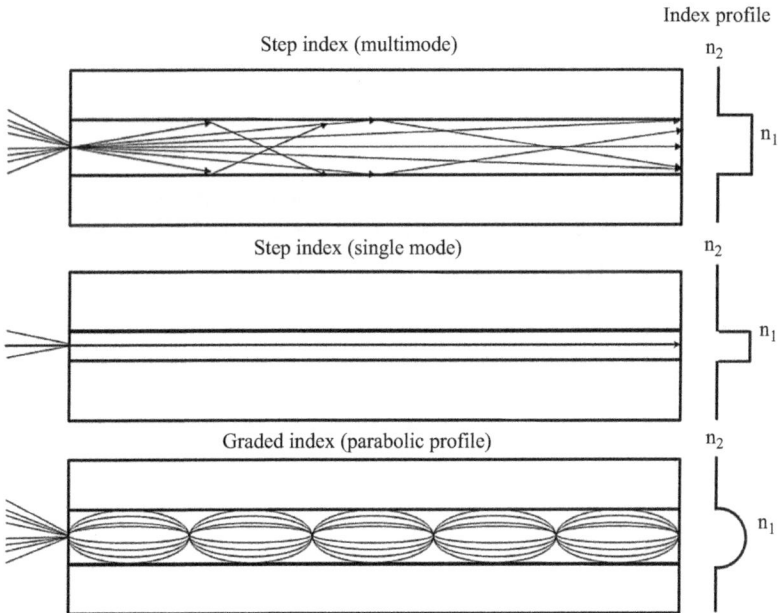

Figure 4.5   Optical fibre modes

Sodnik *et al.* [30] reviewed the technologies of the application of optical communications in platforms mounted in high-altitude locations. These locations are characterised by severe environmental conditions, such as atmospheric disturbance, clouds and turbulences at large that affect the orbit to earth satellite links [31–33]. Hence, by means of compact and light weight terminals, these links have presently been made compatible and fully operable [34].

Shoji *et al.* [35] presented the pioneering experiment showing data transmitted successfully at a speed of 99.7% of the light in vacuum, by using an optical fibre-based WDM data. Li *et al.* [36] reported on a systematic study of the security parameters in-depth analysis for an all-fibre communication network infrastructure for power grid application. By using photonics generation process, Zhang *et al.* [37] determined experimentally an mm-wave generation in the E-band frequency from 2 to 3 GHz.

Finally, to measure the RI, Xu *et al.* [38] proposed a system based on simple FOS. This system consisted of an optical source, an optical coupler and two terminals for fibre sensing, in addition to one mechanical optical switch controllable by applied voltage.

In sum, the fibre optics system is a major constituent in the communication framework, that is why the growth of the fibre optics industry is still continuing increasingly. Because it combines simultaneously low-attenuation capacity and large bandwidth, these assets make it an optimal tool for transmission in the gigabit range and even beyond.

4.3   Ground applications of the fibre Bragg grating sensors

4.3.1   Composite and concrete structures

Contrasted with conventional sensors based on electrical signals, the FBG have many distinguishing advantages including

- their capacity to perform in harsh environmental conditions;
- their small size and flexible shape make them ideal to be embedded into various kinds of composites without affecting the main mechanical properties of the host structures; and
- Highly resistant to corrosive media, especially when they are employed in open environment, like for dams and bridges structures.

These numerous abilities have made them highly suitable devices for many applications for large composite and concrete structures, including impact detection, quality control and health monitoring after construction. Some of the FBG applications in spacecraft, dams and bridges have been detailed by Yokoi *et al.* [39] and are briefly surveyed hereinafter.

4.3.2   Bridges

To address the serious corrosion issue in the large structures, namely the highway bridge, one of the first monitoring demonstrations was the replacement of the tendons made with steel by those fabricated with carbon fibre-based composite materials [39,40]. However, as those composites are still under assessment in concrete structure and did not fully prove their capacity to replace the steel material, a huge activity raised in the controlling and observation of the temperature, strain, mechanical deflection and other climate degradation by means of FBG [40].

4.3.3 Application for dams

Dams are plausibly the gigantic structures in civil engineering domain. Monitoring their mechanical properties during and even after their construction is crucial for their safety, longevity and overall quality. FBG sensors are ideal alternative for health monitoring applications of dams due to their capacity to achieve long-range measurements.

When they are appropriately distributed, FBG are highly interesting because they can ensure measurement over tens of kilometres range with a spatial resolution in the metre scale.

Moreover, the temperature is one of the most important factors to measure in dams, because the frequency of the happening microcracks is directly proportional to the maximum temperature the dams' concrete could undergo during its chemical processing. To monitor this parameter, a Brillouin scattering based on distributed temperature sensor is habitually employed [40].

4.3.4 Application for the mining industry

Mining industry, and particularly the underground galleries activity, is composed of different parts, including miners (or workers), machinery, managers and so on. Such mines are a particular type of environment which requires significant attention to detail since they directly relate to miners' productivity, and more importantly, health and safety. However, negligence in safety aspects may cause damaging of high-quality equipment, hampering production or causing loss of human life in extreme cases. Due to the inconvenience of such environment, high humidity, dust, the possibility of explosions, floods are rending mines hazardous workplaces for the miners and turning the safety issue into a daunting challenge.

Indeed, dynamic measurement of the changes in the displacement and the variation in the load in underground galleries and tunnels is crucial. For this particular mission, the conventional sensors based on electrical signals (including gauges, load cells and so on) that might undergo huge electromagnetic interference (EMI) generated by the various machineries need be replaced by advanced multiplexed FBG sensor. As a matter of fact, a full FBG sensor system can be mounted to measure the long-term static displacement [39].

4.3.5 Application in the electric power industry

In the majority of industrial facilities, measuring the electrical current is of primary and basic importance. Traditionally, the current transformers consisting of iron cores and windings have been used for controlling and supervising electrical current. However, these transformers are known to be heavy and bulky, and the measured signal is subjected to electromagnetic induction noise. It is then inaccurate to measure large current values, particularly for the low-frequency component.

In the 1960s, the Faraday effect was introduced for the first time and consisted of applying an optical current-sensing process [39]. Extensive R&D was then carried out worldwide to develop a reliable current-sensing technology based on this optical principle. During this earlier period of development, sensing elements such as a glass block were employed. Furthermore, a revolutionary technology called optical fibre current sensor (OFCS) that uses the optical fibre as the Faraday-sensing element has been investigated and successfully developed [39].

Like other usages of FOS, owing to their invulnerability against EMI, FBG are highly suitable device for electrical power industry applications. Many applications, including the winding temperature of electrical power transformers and/or loading of power transmission lines, have been successfully measured with the FBG sensor [39], and the efficient use of this sensing technology is continually advancing.

Figure 4.6 is a schematic example showing the configuration of the current sensor device developed for the alternating current measurement which is based on optical fibre using the reflection type [40]. The wideband optical signal is conducted along the optical transmission fibre made with silica (single mode) from the source (e.g. amplified spontaneous emission, 1,550 nm wavelength) to the 'optical box' that is consisting of lens and optical crystals, and where the optical signal is converted to linear polarisation and then launched into the sensing fibre.

At the other end of the fibre, the inserted wavelength is then reflected by a mirror mechanism and transmitted back to the optical box. In the sensor fibre, the Faraday effect is produced by the application of the magnetic field that is induced around the electrical current to be quantified. Then, the optical signal carrying out along the fibre core is introduced into an analyser in the optical box and is then split into two optical signals whose directions of polarisation are orthogonal each other.

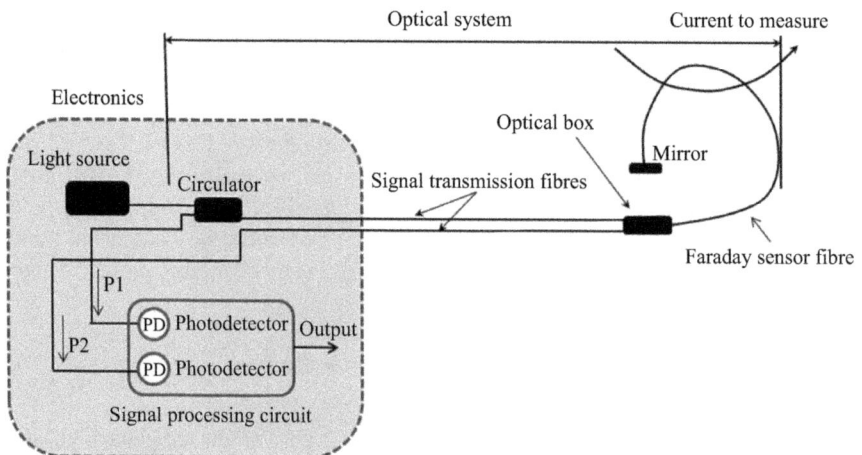

Figure 4.6   Schematic example of reflection type configuration OFCS. P1 and P2 are the two optical signals split right after the optical box

In this configuration, the intensity of the two optical signals is modulated by the Faraday rotation. To ensure a smooth linearity between the light intensity modulation and the Faraday rotation, an angle of 45° optical bias for the polarisation of light is adopted in the optical box by means of a magnetic garnet crystal [40].

Through appropriate photodetectors, the two optical signals are converted to electrical signals. Finally, in the signal processing circuit, the intensity modulation of each light beam is estimated accurately, and the averaging value of the two modulations is reached as the output voltage, proportional to the electrical current to be measured [40].

4.3.6   Application for load monitoring of power transmission lines

During the cold periods, wet snow icing accumulating on high and medium voltage power lines is a serious issue causing failures on electrical supplies. The natural process is a phenomenon in which various variables are contributing and is still not fully understood. In addition to that, the exorbitant mechanical load due to the deposition of heavy snow on electrical power transmission lines may lead to a dramatic accident, in particular, in remote locations (e.g. mountainous areas), where there is no access for easy inspection. Therefore, monitoring the load variation on power line through an online measurement system is highly appropriate for this situation.

This is hence a relevant example for the application of FBG sensors for long-distance remote monitoring under harsh environments condition where the multiplexed FBG system is totally suitable here. As a matter of fact, the load variation can be converted into strain by means of a simple metal plate bonded with an FBG sensor and connected to the power line. Ideally, multiple sensors are appropriate for such an application.

Unlike the WDM which are not appropriate for such a situation because of the limited available bandwidth of the optical signal source and the need of multiple sensors, the time-division multiplexing could be used to increase considerably the multiplexing capacity. High-speed modulation and demodulation are not needed in this scenario since the distance between adjacent FBG sensors is rather large [39].

4.3.7   Application in the petroleum industry and monitoring pipeline

Owing to the numerous inherent advantages of FBG sensors, in particular their safety, ability to work at high temperature and their multiplexing capacity, they constitute an ideal tool for applications in the petroleum industry. Among these assets, the distributed-sensing capability (or multiplexed feature) is of particular interest when the monitoring of different parameters located at different spatial locations is essentially.

In the field of monitoring the pipeline, FOS have become very important tools in the petroleum (oil and gas) industry. FOS technology is now supporting field engineers not only to locate and monitor the oil wells but also to extract the largest possible amount of oil and gas out of them. The wells environmental harsh conditions are really challenging, with temperature over hundreds degrees. The reliability of classical electrical sensors decreases considerably under such environmental factors, and the electrical sensors may even increase the threat of explosion inside the wells. The reason that FOS could offer higher reliability for in well applications is mainly due to their chemically passive behaviour.

Nowadays, the majority of the FOS suitable for petroleum industry available in the market are intrinsic, where the sensing element – as mentioned before – is the fibre itself and the optical signal does not have to exit and re-enter the fibre core.

Hence, after the first demonstration, in 1993, of the FOS based on an optical fibre coupled with a micromachined silicon resonator measuring temperature and pressure in an oil well, many advanced experiments measuring seismic vibration, temperature, pressure and flow have been successfully installed in reservoirs permanently over the past 20 years. Such FOS are widely employed to image, map and characterise the reservoirs in addition to probe and monitor the geophysical properties of rock formations, the processes separating oil, gas and water and so on.

For the pipeline as such, FBG sensors could play a key role to accurately monitor the pressure on the joins and the temperature in the pipe. As the RI of FBG is sensitive to pressure and temperature (pT), by using multiplexing methods described earlier, we can monitor all the length of the pipe by using one fibre only. Figure 4.7 shows a schematic example of the drain of the oil and gas reservoirs where data are provided by a distributed network of flow sensors and pT gauges.

Various methods of leak detection in the pipeline have been deployed so far throughout the world. Some are widely recognised by international standards, while

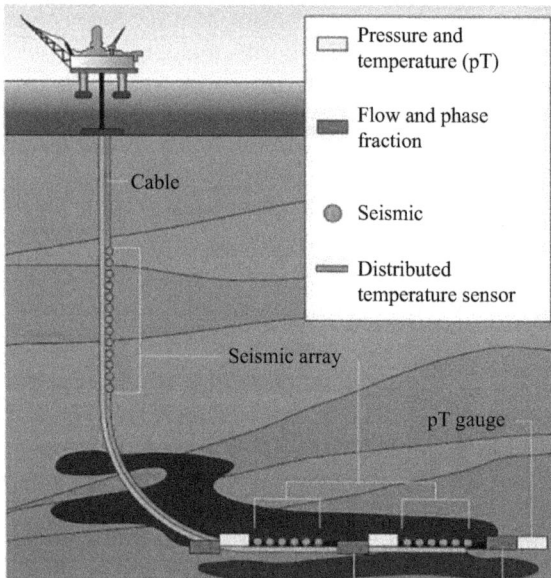

Figure 4.7   Drain of the oil and gas reservoirs. Data provided by distributed flow sensors and pT gauges. © 2008 Nature Publishing Group. Reproduced, with permission, from [41]

others have become more regional or product specific according to their limitations. Distributed fibre optic sensing has been used to detect temperature variations in cryogenic products such as liquid natural gas. This provides a significant delta at the leak point and has become an accepted solution for receiving and export terminals. Natural gas pipelines provide a smaller change at leak and liquid products can soon reach ambient temperatures, especially when combined with the attenuation effects of the ground.

4.3.8   Crack sensors

As the bridges have a finite lifespan, it is inevitable to monitor and assess their shape and health conditions to be able to prevent accidents, mitigate risks and plan maintenance activities accordingly.

Among the most vicious risks are the 'fracture critical bridges' because they have almost no load redundancy. They are hence of particular interest. The health condition of concrete structures could be approached by means of cracking detection and assessment. As a matter of fact, in concrete bridge coating, a crack opening above 150–200 μm will allow the penetration of water and chloride ions, occurring thereby the corrosion of steel reinforcements.

Conventionally, detecting and monitoring cracks occurring into bridges structures has been inspected visually. However, this method is costly, inaccurate and time consuming, in addition to be not reliable. The process called 'structural health monitoring (SHM)' has in recent past appeared as a new topic of engineering, for advancing the evaluation of structural shape. FOS technology has considerably improved the SHM and raised novel alternatives, especially through the use of distributed sensors (i.e. sensing cable). The optical cable is hence sensitive to strain variations generated by damage and cracks at each location on its length. However, these FOS can detect and monitor the crack only if the opening cracking happens in a small location that is already detected *a priori* [42]. An example of a zigzag shape optical fibre encapsulated in the concrete element is schematic in Figure 4.8.

First, the intensity of optical signal along the fibre is measured. For a healthy structure, the backscattered light wave along the optical fibre core is supposed to follow somehow a smooth curve (Figure 4.8) before the cracks formation, where the small signal losses are due only to absorption and scattering of light in the straight portions of the fibre. When the fibre is curved, an additional macrobending loss may occur depending on the radius of curvature [42].

When a crack occurs in the structure, as displayed in Figure 4.8, the portion of the fibre that intersects the crack opening at any angle different from 90° has to bend accordingly to stay continuous. This perturbation in the fibre is brusque and is considered as microbending, resulting in an abrupt drop in the optical signal which is manifested by an intensity loss. This variation in the optical signal is then detected and located through the optical time domain reflectometer.

If the calibration relationship is developed, the crack opening can be directly obtained from the magnitude of the signal loss. This processing does not require

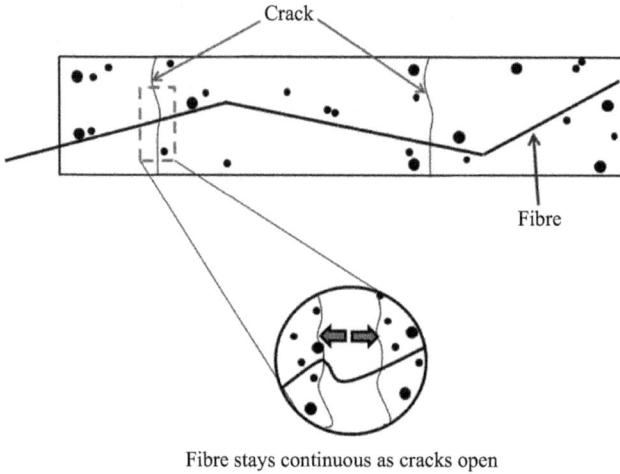

Figure 4.8   Schematic of the principle of the FBG crack detection sensor

prior knowledge of the crack locations. This detection constitutes a serious advancement over existing crack monitoring techniques. In addition to that, several cracks can be detected simultaneously and located with a single fibre.

4.4   Medical applications of the fibre Bragg grating sensors

In addition to their small and miniature size, and due to the fact to be chemically inert, the FOS could be used in various medical applications, especially in the development of minimally invasive surgical tools. Although their market penetration is still slow because of their relatively high cost, a growing number of medical procedures are ameliorated from the leverage that these tiny FOS can offer. A good example came from the neurological surgery, where FOS are used for intracranial pressure monitoring [41]. In addition, such FOS which could measure strain and temperature can be disposed at the tip of instrumented catheters to give useful feedback in surgical procedures. Figure 4.9 shows an example photo of a medical application involving FOS technology [43].

Despite the fact that the majority of medical applications called 'high volume' using FOS necessitate the measurement of physical characteristics, the market of chemical sensing is also growing up. For example, FOS aimed at detecting pH, carbon dioxide and dissolved oxygen are currently available on the market. Moreover, since FOS are immune to EMI, they are perfectly suitable for the field of applications dealing with high electromagnetic topic, e.g. in magnetic resonance imaging systems.

On the other hand, optical fibre biosensors are increasingly attractive in medical applications to detect specific biological species through the measurement of absorbance, reflectance, fluorescence and so on [43]. For example, to detect the change in the concentration of analytes, one of the simplest FOS biosensors

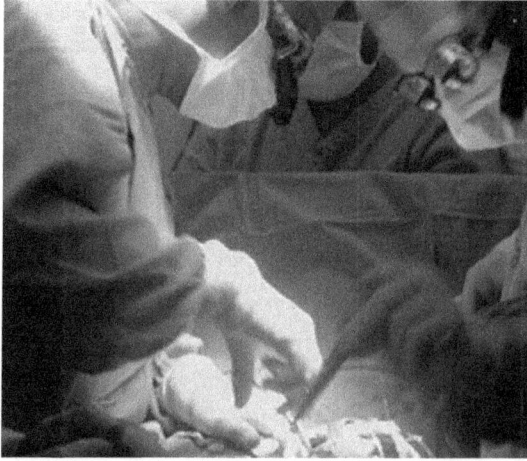

Figure 4.9   FOS have true potential in many medical applications due to their particular flexible and small shape, in particular when dealing with minimally invasive surgical tools. © 2008 Nature Publishing Group. Reproduced, with permission, from [43]

employs the measurement of the optical absorbance which is supposed to occur at particular wavelength. For the pH and oxygen sensing, an optical fibre-based system has been already developed to be deployed in medical application [44,45].

Piunno *et al.* [46,47] reported the pioneering optical fibre deoxyribonucleic acid hybridisation biosensor, while researchers from Virginia Institute and State University developed successfully a microgap fibre multicavity FP sensor for biosensing [48].

Larger parts of medical sensors in the market are based on electrical activation and hence are not suitable for many applications dealing with electromagnetic fields. FOS can overcome this issue since they are somehow dielectric.

Miniature FOS using the modulation of light intensity have been successfully fabricated and already commercialised [49]. In addition, through the unique FBG property of signal multiplexing, it was conceivable to achieve a system of quasi-distributed sensor by using a single-fibre channel.

To measure the heart efficiency, FBG sensors could also be of a particular interest. The idea is based on the flow-directed thermodilution catheter (FDTDC) process. FBG will measure the heart's blood response to an injected cold solution. An FDTDC is inserted into the right atrium of the heart, where the cold solution is injected directly into the heart. FBG could measure the temperature of the blood in the pulmonary artery. By correlating the pulse rate with the read temperature, we could determine with high precision how much heart is pumping blood. This concept of FDTDC with FBG has been already employed as an alternative to the conventional catheter process. To reproduce the effect of blood flow variation due to the change of the blood vessel size, while keeping constant the pump rate, a simple clamp was employed to squeeze the tubing [50].

On the other hand, FBG could be used to detect eye alpha-crystallite aggregation, through the observation and monitoring of the backscattered light intensity from the lens of the eye and by autocorrelation analysis. In fact, the detection is simply characterised by a bimodal distribution of particle size in the backscattered radiation.

4.5   Conclusions

In summary, this chapter reviews the main applications of the fibre optic and FBG sensors in particular. Galvanised by the telecommunication industry, the field of FOS has advanced dramatically during the past few years. As a matter of fact, the technology readiness level associated to R&D in FOS topic has matured considerably, in both academia and industry, and passed from proof of principle demonstration to the production of test instruments and the engineering of final prototypes.

Researchers are currently focusing on issues related to the application of optical fibres in harsh environments. FOS will particularly play a key role in the structural monitoring context where the R&D effort would be mature.

4.6   Annex

Table 4.1 summarises the main telecommunications applications of the FBG sensors together with other applications and the corresponding relevant references.

Table 4.1   Main telecommunications applications of the optical fibre and FBG sensors

Telecommunications applications	
Application	**Reference**
Dispersion compensation	[50]
Wavelength selective devices	[51]
Band-rejection filters, long-period gratings	[52]
Fibre taps	[53]
Fibre erbium amplifiers	[54]
Network monitoring and optical fibre identification	[55]
Cascaded Raman amplification at 1.3 μm	[56]
Fibre lasers	[57]
Semiconductor lasers with external Bragg grating reflector	[58]
Other applications	
Optical fibre mode converters; spatial mode converters, polarization mode converters	[59]
Grating-based sensors	[60]
Optical single processing; delay line for phased array antennas, fibre grating compressor	[61]
Nonlinear effects in fibre Bragg gratings; optical switching, optoelectronic devices, wavelength conversion devices	[62]
Optical storage; holographic storage, direct writing	[63]

References

[1] F.T.S. Yu and S. Yin in *Fiber Optic Sensors*, Marcel Dekker Inc., New York, NY, USA, 2002.

[2] D. Kersey, M.A. Davis, H.J. Patrick, *et al.*, *Journal of Lightwave Technology*, 1997, **15**, 8, 1442.

[3] C.E. Lee, H.F. Taylor, A.M. Markus and E. Udd, *Optics Letters*, 1989, **14**, 21, 1225.

[4] B. Dong, L. Wei and D.P. Zhou, *Applied Optics*, 2009, **48**, 33, 6466.

[5] M.A. Davis, D.G. Bellemore, T.A. Berkoff and A.D. Kersey, *Proceedings of SPIE*, 1995, **2446**, 227.

[6] A.D. Kersey, M.A. Davis and D. Bellemore, *Proceedings of SPIE*, 1995, **2456**, 262.

[7] M.G. Xu, H. Geiger, J.L. Archambault, L. Reekie and J.P. Dakin, *Electronics Letters*, 1993, **29**, 17, 1510.

[8] M.A. Davis and A.D. Kersey, *Electronics Letters*, 1995, **31**, 10, 822.

[9] A.H. Hartog, *Journal of Lightwave Technology*, 1983, **1**, 3, 498.

[10] J.P. Dakin, D.J. Pratt, G.W. Bibby and J.N. Ross, *Electronics Letters*, 1985, **21**, 13, 569.

[11] X. Bao, D.J. Webb and D.A. Jackson, *Optics Letters*, 1993, **18**, 7, 552.

[12] O. Frazao, J.M. Baptista and J.L. Santos, *Sensors*, 2007, **7**, 11, 2970.

[13] J.C. Juarez and H.F. Taylor, *Optics Letters*, 2005, **30**, 24, 3284.

[14] J.C. Juarez and H.F. Taylor, *Applied Optics*, 2007, **46**, 11, 1968.

[15] X. Bao, C. Zhang, W. Li, M. Eisa, S. El-Gamal and B. Benmokrane, *Smart Materials and Structures*, 2008, **17**, 1, 015003.

[16] R.J. Sanferrare, *AT&T Technical Journal*, 1987, **66**, 95.

[17] D. Gloge, A. Albanese, C.A. Burrus, *et al.*, *The Bell System Technical Journal*, 1980, **59**, 1365.

[18] O. Lopez, A. Amy-Klein, C. Daussy, *et al.*, *The European Physical Journal D*, 2008, **48**, 1, 35.

[19] T. Mizumoto, H. Chihara, N. Tokui and Y. Naito, *Electronics Letters*, 1990, **26**, 3, 199.

[20] T. Otani, K. Goto, H. Abe, M. Tanaka, H. Yamamoto and H. Wakabayashi, *Electronics Letters*, 1995, **31**, 380.

[21] G.A. Thomas, B.L. Shraiman, P.F. Glodis and M.J. Stephan, *Nature*, 2000, **404**, 262.

[22] Y. Shoji and T. Mizumoto, *Optics Express*, 2007, **15**, 13446.

[23] C.E. Shannon, *Bell System Technical Journal*, 1948, **27**, 379.

[24] M. Noshada and A. Rostami, *Optik*, 2012, **123**, 758.

[25] X. Wang and K. Kitayama, *Journal of Lightwave Technology*, 2004, **22**, 10, 2226.

[26] Y. Shoji, I-W. Hsieh, R.M. Osgood and T. Mizumoto, *Journal of Lightwave Technology*, 2007, **25**, 10, 3108.

[27] S-H. Kim, R. Takei, Y. Shoji and T. Mizumoto, *Optics Express*, 2009, **17**, 11267.

[28] M. Amemiya, M. Imae, Y. Fujii, T. Suzuyama, F. Hong and M. Takamoto, *IEEE Transaction on Instrumentation and Measurement*, 2010, **59**, 3.

[29] G. Grosche, B. Lipphardt, H. Schnatz, *The European Physical Journal D*, 2008, **48**, 27.

[30] Z. Sodnik, B. Furch and H. Lutz, *IEEE Journal of Selected Topics in Quantum Electronics*, 2010, **16**, 1051.

[31] L.C. Andrews, R.L. Phillips, C.Y. Hopen and M.A. Al-Habash, *Journal of Optical Society of America A*, 1999, **16**, 1417.

[32] N.J. Collela, J.N. Martin and I.F. Akyildiz, *IEEE Communications Magazine*, 2000, **38**, 6, 142.

[33] G. Avdikos, G. Papadakis and N. Dimitriou, *Proceeding of the 10th Signal Processing for Space Communications*, Rhodes, Greece, 6–8 October 2008.

[34] T. Jono, Y. Takayama, K. Shiratama, *et al.*, *Proceedings of SPIE*, 2007, **6457**, 645702–1.

[35] Y. Shoji, T. Mizumoto, H. Yokoi, I-W. Hsieh and R.M. Osgood, *Applied Physics Letters*, 2008, **92**, 071117.

[36] X. Li, J. Yu, Z. Dong and N. Chi, *IEEE Photonics Journal*, 2013, **5**, 1, 7900107.

[37] L. Zhang, B. Liu, X. Xin and Y. Wang, *IEEE Photonics Technology Letters*, 2013, **25**, 4.

[38] W. Xu, X.G. Huang and J.S. Pan, *IEEE Sensors Journal*, 2013, **13**, 5.

[39] H. Yokoi, T. Mizumoto, N. Shinjo, N. Futakuchi and Y. Nakano, *Applied Optics*, 2000, **39**, 33, 6158.

[40] K. Kurosawa, *Photonic Sensors*, 2014, **4**, 12.

[41] H. Nakstad and J.T. Kringlebotn, *Nature Photonics*, 2008, **2**, 147.

[42] J.R. Casas and P.J.S. Cruz, *Journal of Bridge Engineering*, 2003, **8**, 6, 362.

[43] E. Pinet, *Nature Photonics*, 2008, **2**, 150.

[44] R. Wolthuis, D. McCrae, E. Saaski, J. Hartl and G. Mitchell, *IEEE Transactions on Biomedical Engineering*, 1992, **39**, 5, 531.

[45] R.A. Wolthuis, D. McCrae, J.C. Hartl, *et al.*, *IEEE Transactions on Biomedical Engineering*, 1992, **39**, 2, 185.

[46] P.A.E. Piunno, U.J. Krull, R.H.E. Hudson, M.J. Damha and H. Cohen, *Analytica Chimica Acta*, 1994, **288**, 3, 205.

[47] P.A.E. Piunno, U.J. Krull, R.H.E. Hudson, M.J. Damha and H. Cohen, *Analytical Chemistry*, 1995, **84**, 4, 1854.

[48] Y. Zhang, X. Chen, Y. Wang, K.L. Cooper and A. Wang, *Journal of Lightwave Technology*, 2007, **25**, 7, 1797.

[49] K.O. Hill, *Applied Optics*, 1974, **13**, 1853.

[50] D. Gatti, G. Galzerano, D. Janner, S. Longhi and P. Laporta, *Optics Express*, 2008, **46**, 1945.

[51] K.O. Hill, B. Malo, F. Bilodeau, S. Thériault, D.C. Johnson and J. Albert, *Optics Letters*, 1995, **20**, 143.

[52] A.M. Vengsarkar, P.J. Lemaire, J.B. Judkins, V. Bhatia, T. Erdogan and J.E. Sipe, *Journal of Lightwave Technology*, 1996, **14**, 1, 58.

[53] A.M. Vengsarkar, P.J. Lemaire, J.B. Judkins, V. Bhatia, T. Erdogan and J.E. Sipe, *Proceedings of the Optical Fiber Communication '95*, San Diego, *CA*, USA, 1995.

[54] K.O. Hill, B. Malo, K.A. Vineberg, F. Bilodeau, D.C. Johnson and I. Skinner, *Electronics Letters*, 1990, **26**, 1270.

[55] R. Kashyap, R. Wyatt and P.F. McKee, *Electronics Letters*, 1993, **29**, 1025.

[56] K.O. Hill and G. Meltz, *Journal of Lightwave Technology*, 1997, **15**, 1263.

[57] G.A. Ball, W.W. Morey and J.P. Waters, *Electronics Letters*, 1990, **26**, 1829.

[58] G.A. Ball and W.W. Morey, *IEEE Photonics Technology Letters*, 1991, **3**, 1077.

[59] P.A. Morton, V. Mizrahi, S.G. Kosinski, *et al.*, *Electronics Letters*, 1992, **28**, 561.

[60] A.D. Kersey, *Optical Fiber Technology*, 1996, **2**, 291.

[61] G. Meltz, W.W. Morey, W.H. Glenn and J.D. Farina, *Proceeding of the Optical Fiber Sensors'88*, New Orleans, LA, USA, 1988.

[62] C.M. de Sterke, N.G.R. Broderick, B.J. Eggleton and M.J. Steel, *Optical Fiber Technology*, 1996, **2**, 253.

[63] T. Erdogan, A. Partovi, V. Mizrahi, *et al.*, *Applied Optics*, 1995, **34**, 6738.

Chapter 5

Main fibre Bragg grating fabrication processes

The idea behind the fabrication of Bragg gratings inside the core of an optical fibre has grown since the establishment of the photosensitivity phenomenon. The capacity to imprint Bragg gratings within the core in these photosensitive fibres has reformed the field of telecommunications and fibre optic sensor(s) (FOS) bed technology. Over the last decades, the number of research and development investigations for both fundamental phenomena and application of the fibre gratings has expanded seriously. In this chapter, we introduce and review the technology of Bragg gratings in optical fibres. We detail the aspect of photosensitivity in optical fibres, the properties of Bragg gratings, and the main developments in devices and applications. The most dominant fabrication techniques, including interferometric, phase mask and point-by-point are developed and their respective advantages/disadvantages discussed accordingly.

5.1 Introduction

The deployment of fibre optics in telecommunications has revolutionised this field rending good quality, high capacity and long-distance telephone communication possible. Over the past 40 years, advancements and reshaped fibre optic technology has continued, until nowadays, where optical fibres have become somehow synonym of 'telecommunication'. However, in addition to applications in telecommunications, optical fibres are increasingly employed in the fast-growing fields of fibres lasers, amplifiers and sensors. Even though huge advancements have been recorded in the field of optical fibres in general, fundamental optical components including mirrors, optical filters, and complexes reflectors have been a true challenge to be integrated with optical fibres geometry.

However, recently, all these technological issues have become more affordable with the capacity to alter the refractive index (RI) in the core of a single-mode optical fibre through ultraviolet (UV) irradiation. The ability of photosensitivity in the optical fibres permits the fabrication of phase structures or gratings in the core of fibres which could be generated by permanently varying the RI in a periodic diagram along the core of the fibre. A cyclic modulation of the RI in the fibre core plays the role of a selective mirror for the optical signal wave that meets the Bragg condition, forming thereby what is commonly known today as a fibre Bragg grating (FBG).

The characteristics of the grating including its length and period, and the amplitude of the RI modulation, all determine whether the grating has a high or low reflectivity over a large or small range of wavelengths. For that reason, these parameters determine the role of the Bragg grating, if it can act as wavelength division multiplexer in telecommunications, e.g. a narrowband high-reflectance mirror in laser; sensor applications and filter of wavelength selection against undesired laser frequencies in fibre amplifiers.

The first observation of the RI variation by Hill *et al.* [1,2] was recorded in 1978 in germanosilica fibres. Authors outlined the ability to write a stable and invariable grating pattern in the core of the optical fibres by using a blue Ar ion laser line emitting at 488 nm. This specific grating had the disadvantage of offering a very weak RI modulation, estimated of about 1,026 nm at the writing wavelength, stemming in a narrowband reflection.

After the pioneering work on the photosensitivity discovery in the optical fibres by Hill *et al.* [1], no advancements were recorded for the next 10 years. This was attributed mainly to the limitations technology of the writing technique at that period. Meltz *et al.* [3] have then renewed this interest by demonstrating the side writing principle. Following this work, many groups over the world successfully accomplished breakthroughs in writing high-quality gratings inside the core of optical fibres, through many technological processes such as the phase mask, point-by-point exposure to UV radiation and interferometry. Gratings offering a large range of bandwidth and reflectivity can be processed from a few nanoseconds (which is the duration of the laser pulse) to a few minutes timescale, depending on the required performance. In addition, these gratings offer very low optical losses and can be inscribed gently into the optical fibre core at the desired location with high spatial accuracy, offering, e.g., a very narrow bandwidth wavelength selection.

The sensing-based optical fibre (or FOS) is a topic that has been linked with the Bragg gratings concept from the beginning. The discovery of the photosensitivity has led to new field of devices [4–6], especially in sensing and telecommunication applications based on innovative and new Bragg gratings structures. As a matter of fact, bandpass filters based on fibre Fabry–Pérot (FP) Bragg gratings, blazed gratings for mode converters and chirped gratings for dispersion compensation are becoming routine applications.

5.2   Fundamentals of the photosensitivity in optical fibres

At the Communication Research Center in Canada, a Canadian laboratory, Hill *et al.* [1,2] revealed in 1978 the photosensitivity in germanium-doped silica fibre. This discovery happened somehow by chance during an experiment which was primarily settled to investigate the non-linear effects in a particularly designed optical fibre. A visible optical signal generated by argon ion (Ar^+) blue laser was sent into the core of the optical fibre. Upon extended irradiation, the attenuation of the optical fibre was observed to be more pronounced. Based on this, it was established that the intensity of the back-reflected optical signal from the fibre increased significantly with respect to the exposure time. This enhancement of the light reflectivity was the consequence of a photoinduced RI grating in the fibre. This non-linear photorefractive effect occurring in the optical fibres core was called 'fibre photosensitivity' and is a permanent (non-reversible) phenomenon.

From physics point of view, in the Hill's experiment [1], the blue 488 nm laser beam was launched into the core of a particularly designed optical fibre. This laser beam interfered then with the Fresnel reflected beam to form a weak pattern of standing wave intensity. The high-intensity points modified the RI value in the photosensitive fibre core irreversibly. Like so, a perturbation in the RI having the same spatial regularity as the interference pattern is formed, having a length defined solely by the coherence length of the inscribing radiation. This RI grating plays the role of a distributed reflector that conjugates the progressive propagating signal to the opposed propagating optical beams. This coupling improves the amplitude of the back-reflected optical signal and hence increases the intensity of the interference pattern, which in turn increased the RI at the high-intensity points. This procedure is repeated until the reflectivity of the grating reach eventually a saturation level. The forming gratings are called self-organised or self-induced gratings as they are formed automatically without human intervention. This work was performed on a particularly configured optical fibres furnished by Bell Northern Research Fibre Laboratory. These specifically designed fibres are heavily doped with Ge and are distinguished by their small core diameter.

During the decade between Hill's experiment and 1981, only few research works were pursued on fibre photosensitivity and were located only in Canada at the Bell Northern Research Fibre Laboratory. In 1981, Lam and Garside [7] demonstrated that the RI has a magnitude variation which depended on the square of the laser power (i.e. fabricating Ar^+ blue laser with a wavelength of 488 nm). The hypothesis of a two-photon mechanism was discussed as the possible phenomenon of the RI variation. Unfortunately, during that period of time, the lack of international interest in fibre photosensitivity was associated to the effect being considered as a mechanism characterising this special kind of fibre only. A decade later, Stone [8] demonstrated otherwise. In fact, he successfully and repeatedly noticed a photosensitivity phenomenon occurring in various different Ge-doped fibres.

In the optical fibres structures, the self-organised Bragg gratings act as narrowband reflection filters. Hill *et al.* [1] employed this kind of gratings rather than the classical output reflector of an Ar^+ laser. Doing so, they were capable to realise a stable continuous wave (CW) oscillation at 488 nm wavelength (Figure 5.1). This achievement illustrated the pioneer recorded demonstration of a dispersed assessment CW oscillation of a visible gas laser. The output reflector was detached from the laser head followed by coupling the laser beam to the optical fibre core that is incorporating the reflection filter through the lens of an optical microscope objective. Despite the fact that the ascertainment of the photosensitivity in the form of photoinduced RI variations has had a central role in the development of optical fibre technology, devices such as the self-induced gratings were not feasible, due to the fact that the Bragg resonance wavelength was linked to the 488 Ar^+ wavelength. Modest tunability of the resonance frequency could be realisable during the

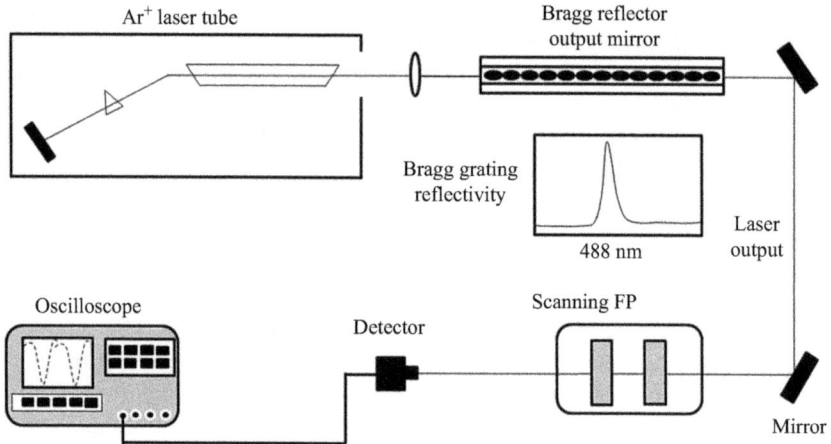

Figure 5.1   Detailed schematic of a distributed feedback (DFB) CW 488 nm Ar⁺
            laser beam

Bragg grating writing process through the exercise of longitudinal strain. However, this strain-based processing method is not suitable for the formation of Bragg gratings having a resonance frequency located in the infrared (IR) range, which is highly in demand for optical communications. In addition, in the blue–green spectral frequency range, these gratings are inherently unstable due to the permanent photosensitivity of the fibre.

As a consequence, the grating will constantly evolve while it is used as a Bragg reflector. On the other hand, the grating could also totally disappear if it is irradiated with different blue–green light frequency [9]. However, owing to the two-photon process, an additional issue related to the small photoinduced RI variation could also happen with the self-induced gratings. Accordingly, to realise a detectable reflectivity capability, long gratings are more appropriate, which, in turn, rend these types of gratings impracticable for localised-sensing applications.

At a spectrum range of 240–250 nm, Meltz *et al.* [3] showed in 1989 the arising of a strong RI variation when a Ge-doped fibre was irradiated with UV light close to the absorption peak of a Ge-related defect. Figure 5.2 displays an oxygen-deficient germanium defect – localised at 240 nm [10–12] – which is believed to be the source of the photosensitivity property in germane silica fibres.

Hand and Russell [13] developed a theoretical model based on the Kramer–Kronig correlation to interpret the RI variation by relating it to the absorption variation. This model suggests the fracturing of the GeO defect producing a GeE' centre by releasing one electron that is free to move within the fibre glass matrix until it is re-trapped.

This photosensitivity phenomenon is not limited to the germanosilica fibres. In fact, fibres doped with cerium, europium and erbium/germanium [14–16] manifest different levels of photosensitivity, but none is equal to the germanium-based one intensity. The germanium–boron co-doping fibres show a large index modulation

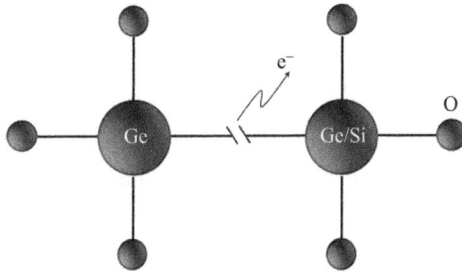

Figure 5.2 Oxygen-deficient germanium defects which are believed to be the source of the photosensitivity effect in germanium-doped silica

of about 10^{23} [17]. Also, fluorozirconate fibre doped with Ce/Er manifests a photosensitivity, where Bragg gratings were written by using a UV 246 nm light [18].

5.3 Defects in germanosilica optical fibres

The variation of RI in the germanosilica fibre is activated by a single photon at 240 nm wavelength. This triggering radiation frequency is definitely below the band gap located at 146 nm. This highly suggests that the point defects in the ideal tetrahedral network in the glass matrix could be responsible for this mechanism.

The field of studying the point defects in the optical fibres has captivated consideration primarily to find solution to avoid the unwanted absorption bands associated with them and their associated transmission losses.

These defects are also named 'colour centres' owing to their high absorption. Typically, these defects occur during the fibre writing and ionising-radiation processes [19,20]. A huge research effort has been oriented towards decreasing as much as possible the formation of these defects [21]. Nonetheless, since their application to FBGs, the role and responsibility of the defects have changed completely.

For the modified chemical vapour deposition (MCVD) deposited Ge–silica optical fibre, obtained experimental results suggested that the Ge–Si *wrong bonds* could be the source of the photosensitivity property [22]. Despite the fact that the Ge–Si wrong bonds are known to provoke the different processes triggering the RI variation on the glass material, it is worth noting here that the clear responsibility of the Ge–Si bonds should be highlighted accurately since it may not be the triggering mechanism alone but probably the most efficient known so far. The absorption peak of the Ge–Si wrong bond is located at 240 nm [22]. As a direct consequence of the UV light processing, a blanching of this absorption band occurs [23] which has for effect to generate more defects and lead to the evolution of additional absorption bands and so on.

5.3.1 Intensification of the photosensitivity property in silica optical fibres

The photosensitivity in the optical fibres core should be understood in terms of the rate of the RI change and/or variation following a particular UV light irradiation.

Considering the establishment of this phenomenon in the germanosilica fibres and the first experimental establishment of the grating formation, huge research efforts were conducted to achieve the intensification of the photosensitivity property. Primary results have shown that high Ge-doped optical fibres presented the highest photosensitivity under reduced oxidising conditions. More recently, many chemical treatment techniques have proven a clear improvement of the germanosilica fibres photosensitivity, including flame brushing, hydrogenation or hydrogen loading and flame brushing treatment.

Moreover, it has been found that irradiation with a 193 nm ArF excimer laser could present an interesting alternative process for writing and fabricating intense Bragg gratings in the optical fibres without recourse to the hydrogen loading. This technique could be seen as an enhancement of the photosensitive to the conventional range of the irradiation band, namely the 240–250 nm for writing the Bragg gratings in the germanosilica fibres.

5.3.1.1   Hydrogenation of optical fibre (H$_2$ loading)

In germanosilica optical fibres, one of the most efficient methods to improve the UV photosensitivity is through hydrogen loading, also called hydrogenation [24]. It is achieved simply by means of the diffusion of hydrogen molecules into the core of the optical fibre carried out at high temperatures and high pressures. This process, which improves considerably the photosensitivity level, is followed by a permanent variation of the RI into the core of the optical fibre (up to 0.01) through the UV exposure. This variation of the RI is equal to the difference of the index between the core and the fibre cladding. It is worth noting at this level that the variation of the optical fibre or the waveguide photosensitivity following the hydrogenation is not an irreversible effect, because as the hydrogen molecules diffuse out, the photosensitivity comes back to its initial value.

5.3.1.2   Flame brushing

The flame brushing is a straightforward processing occurring in the germanosilica optical fibre to improve its photosensitivity [25]. Typically, we use hydrogen-rich flame (with some amount of oxygen) to brush out repeatedly the area of the optical fibres or waveguide to be photosensitised. This process generates temperatures approaching 1,700°C. The procedure of photosensitisation takes about 20 min to be completed. At this level of temperature, the hydrogen molecules diffuse inside the core of the optical fibre rather promptly and react with the germanosilica glass to produce centres of Ge–O deficiency. These centres produce a substantial absorption band in the germano-doped fibre core, located at 240 nm, transforming the core material highly photosensitive in such a way that the UV radiation could introduce a significant variation in its RI.

5.3.1.3   Boron co-doping

When operating the boron as a co-dopant in germane silica fibre, the UV photosensitivity has been found to increase significantly in the optical fibres [26], comparatively with highly doped germanium fibres alone. Moreover, RI saturation changes were obtained faster and easier than for any other optical fibres, meaning that additional mechanism could be involved in the boron co-doped fibre.

5.3.1.4 Argon fluoride excimer ultraviolet laser radiation

In the optical fibres, the photosensitivity phenomenon and the associated imprinting of Bragg gratings have been always linked to the bleaching of the UV absorption band occurring around 245 nm (5.0 eV). Lately, it has been successfully determined that Bragg grating could be written in optical fibres employed in telecommunication by using an argon fluoride (ArF) excimer vacuum UV laser with a wavelength located at 193 nm [27,28]. As a matter of fact, Albert *et al.* [27] successfully fabricated FBGs by means of a krypton fluoride (KrF) ~248 nm and ArF ~193 nm excimer lasers through a mechanical phase mask. At similar excitation parameters, the Bragg gratings inscribed by the ArF 193 nm irradiation was found to show a higher light reflectivity than those fabricated with the KrF 248 nm. It is evident from these findings that the ArF excimer laser irradiation offers an efficient source for fabricating RI patterns in the germanosilica optical fibres. One of the main advantages behind using shorter wavelength laser or light sources in writing the Bragg gratings is the opportunity to have a high spatial resolution especially in the diffraction-limited methods, like the point-by-point technique.

5.3.2 Mechanism behind the photoinduced refractive index variation

Although the Bragg gratings have been successfully fabricated in various types of optical fibres using different processes, the exact mechanism behind the RI variation is not yet completely understood. Numbers of different mechanisms have been established for these photoinduced RI variations where the exclusively prevalent element in all these assumptions is that the vacancy defects in the germanium–oxygen, namely Ge–Si and/or Ge–Ge (known as 'wrong bonds'), are behind the RI variation.

All along the MCVD fabrication process, and especially at the same time as the high-temperature gas-phase oxidation step, GeO_2 is dissociated to GeO due to its higher chemical stability at high temperatures. This GeO species, when integrated into the glass material of the optical fibre, could manifest themselves in the form of oxygen vacancy Ge–Si and Ge–Ge 'wrong bonds'. As the germane silica core optical fibres which are heavily doped are photosensitive to the UV exposure located habitually in the interval of 240–250 nm, these 'wrong bonds' defects have been precisely correlated to the models of photoinduced RI variations in each of the established mechanisms. Despite the fact that there is some experimental evidences for the efficacy of some of the suggested standards, there are still contradictory results about their quantified contribution to the measured RI variation induced. It is then proposed the possible involvement of more than one process in this photoinduced RI variation phenomenon, and consequently, to the progress of the grating formation. Following are some of the main established models.

5.3.2.1   The model of colour centre

The colour centre standard supposes that variation in the photoinduced absorption spectrum engender the variation in the RI which is expressed by the Kramers–Kronig law.

The relationship of Kramers–Kronig is expressed by [29]

$$\varepsilon_r(\lambda) = 1 + \int \frac{\varepsilon_i(\lambda)}{\lambda' - \lambda} \partial \lambda' \tag{5.1}$$

where ∂ is the symbol of derivation, ε is the permittivity (or dielectric constant), ε_r is the relative permittivity is and ε_i relative permittivity i to be integrated over all λ. λ is the wavelength and it describes the real and imaginary parts of the dielectric constant, namely

$$\varepsilon = \varepsilon_p + i\varepsilon_l = (n + ik)^2 \tag{5.2}$$

where ε is the permittivity (or dielectric constant), ε_p is the real part of the permittivity, ε_l is the imaginary part of the permittivity, i is the imaginary number, $i^2 = -1$, where n is the RI and k is the absorption index.

The Kramers–Kronig correlation would have been born from the origin reason for the dielectric response and establishes that the RI variation occurred in the range of IR/visible light by the photoinduced processing is the consequence of the variation in the absorption spectrum of the glass material in the UV/far-UV light bands region.

In this paradigm established by Hand and Russell [13], the UV irradiation affects the properties of the glass material and institutes new colour centres (electronic transitions of defects). The highlighting assumption of this model is the fact that the photosensitive effect results from confined electronic excitations of defects. The 'wrong bond' defects, which at the beginning are absorbing the light, are now transformed to defects that are more polarisable because their electronic transitions happen at longer wavelengths and/or have powerful transitions.

5.3.2.2   Dipole model

The dipole model is established on the arrangement of deep-seated regular space-charge electric fields through the photoexcitation of 'wrong bond' defects (Figure 5.3). The photoionisation of the Ge–O, Ge–Si and/or Ge–Ge deficient centres creates both free electrons and GeE hole centres positively charged. As the free electrons diffuse distantly and undergo trapping at the nearby sites Ge(1) and Ge(2) to structure new negatively charged Ge(1)$^-$ and Ge(2)$^-$ electron traps, respectively [30], they result, together with the GeE hole centres traps, in a several angstroms spacing electric dipoles. Each electric dipole will yield an electric field with a static direct current 'dc' polarisation which induces local RI variation that is directly proportional to square of the electrical field of the dipole source E^2 *via* the dc Kerr effect. All along, the Bragg grating inscription mechanism when the fibre core is UV irradiated, the free electrons will diffuse from the high intensity zones and are captured by defects in the low intensity regions. This equilibrium owing to the charges redistribution in the optical fibre will generate regular space-charge electric fields.

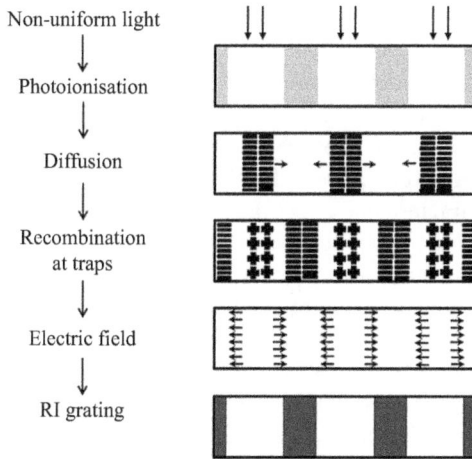

Figure 5.3 Reaction of the photorefractive material to a sinusoidal spatial light pattern

The periodic RI variation is directly proportional to the non-linear coefficient $\chi^{(3)}E^2$, where $\chi^{(3)}$ is the third-order non-linear coefficient and E is the electric field of the dipole source.

5.3.2.3 Compaction (or compression) model

As indicated by its name, the compaction model refers to the change in the material density as a consequence of the laser irradiation, which in turns induces a change in the RI. Indeed, low-intensity illumination by excimer KrF laser (248 nm) well down the breakdown threshold has been found to reversibly generate thermally a linear compression in the amorphous silica occasioning an RI change. Fiori *et al.* [31] irradiated thin-film amorphous silicon dioxide grown on silicon wafers by a KrF excimer laser. As a matter of fact, decrease of about 15% in the film thickness occurred at an irradiation cumulus dose of 2,000 J/cm², leading thereby to an evolution of the RI during the laser irradiation process. Then, after a thermal annealing at 950°C for 60 min in an ultra-high vacuum of 10^{-6} Torr, the material compression evanesced and the initial film thickness is retrieved together with the original RI value. However, it is worth noting here that a continual cumulation of laser irradiation above this reversible compression dose forces an irreversible compression regime until the film was fully etched out (i.e. zero thickness) after a total accumulated dose of 17,000 J/cm². This concept of compression model is somehow novel and so far has not been yet in-depth investigated, and despite the fact that there is some research works published on the compaction in Ge–silica optical fibres, this area is still poorly understood.

5.3.2.4 Stress-relief model

This model [29] is based on the assumption that the RI variation is the consequence of the mitigation of inherent thermoelastic stresses in the core of the fibre. In the germanosilica fibre, the core is under tension because of the existing gradient in the thermal expansion coefficient between the fibre core and the cladding; since the glass is cooled down, the fictive temperature over the fibre writing.

Through the stress-optic effect, it is well established that that tension diminishes the RI value *via* the effect of the stress-optic, and hence, it is anticipated that the stress relief will promote the RI value. Throughout the UV exposure, the 'wrong bonds' will fracture and boost the relaxation in the glass material which is under tension, reducing thereby the thermal stresses in the fibre core. In this stress-relief mechanism, the variation of the RI results from the stress relief in the glass fibre core. This model is triggered by the breakage of the 'wrong bonds' under UV light irradiation. However, although there is huge number of available 'wrong bonds' that are potentially breakable in the Ge–silica core fibres, this is not the same case for pure silica core fibres, which are also not photosensitive to the UV exposure. The stress-relief model, like the compaction mechanism, is as well somehow new and needs in-depth research to establish its industrial validity or not.

5.4   Processes for the Bragg gratings inscription in the optical fibres

5.4.1   *Externally written Bragg gratings in optical fibres*

The necessity of a microscale periodic pattern to fabricate the Bragg gratings in optical fibres rends the stability of the inscription a true challenge to face for all the writing processes. Nowadays, only a very few externally inscription fabrication processes exist, including the interferometric technology, the phase-mask process and the point-by-point technique.

5.4.1.1   Interferometric fabrication technique

Meltz *et al.* [3] achieved the first external fabricating process to write the Bragg gratings in a photosensitive optical fibre [3] by using an interferometer with the particularity to divide the UV laser into two distinguished beams. The two incoming lights are then recombined to generate the interference fringe pattern, irradiating the photosensitive fibre core and inducing an RI change. As a matter of fact, in optical fibres, Bragg gratings have been inscribed through two main interferometers, namely, amplitude and wavefront splitting which will be detailed in the next section.

Amplitude-splitting interferometer

In this mode called amplitude-splitting interferometer, the UV writing laser is first split into two beams of equal intensity. These beams will be later recombined after traveling through two different optical paths. This creates then an interference pattern at the core of the photosensitive optical fibre. To match accurately the interfering beams to the core of the optical fibre, cylindrical lenses are put into the interferometer with a normal angle (90°). The period of the interference fringe pattern is identical to that of the Bragg grating 'Λ' and is function of the exposure wavelength λ_w, and the half-angle between the intersecting UV laser beams φ (Figure 5.4).

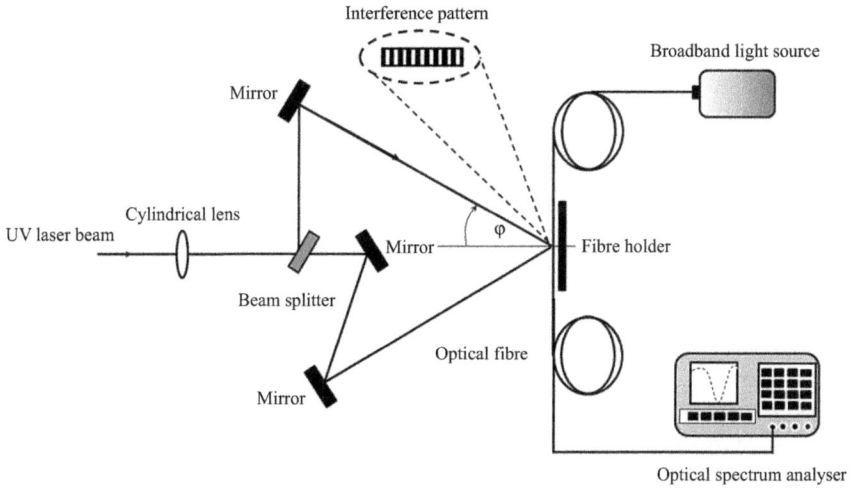

Figure 5.4 Experimental set-up for the inscription of Bragg gratings network in the optical fibre by UV irradiation interferometer method

The Bragg grating period 'Λ' is given by

$$\Lambda = \frac{\lambda_w}{2 \sin \varphi} \tag{5.3}$$

where λ_w is the wavelength of the UV light, and φ is the half-angle between the intersection of the two UV laser beams. Even though the interference pattern is produced into the core of the optical fibre within glass material, its period is identical to what it would be if the laser beam was interfering in air media. This is the consequence of the optical refraction of the light beams that are coupled with the wavelength abridgment as they enter the glass of the fibre core.

The Bragg law, $\lambda_B = 2n\Lambda$, suggests that the Bragg resonance wavelength in the core of the optical fibre, namely λ_B, is two time the product of the RI of the core n and the period of the grating. Hence, the wavelength of the Bragg grating resonance can be expressed in terms of the UV writing wavelength and the half-angle between intersecting UV laser beams as

$$\lambda_B = \frac{n\lambda_w}{\sin \varphi} \tag{5.4}$$

From (5.4), one can clearly notice that the Bragg grating wavelength can be altered either by varying λ_w and/or by changing φ.

Any movement of the beam splitter, interferometer and the position of the various mirrors, even as small as submicrons, will result in the drift of the fringe pattern, dismissing the grating.

Besides, because of the long lengths separating the optical paths inside the interferometers, the air currents will impact the RI locally, and affect the good formation of a stable fringe pattern. Further to the raised shortcomings, gratings of high quality can fabricate only a laser light source that has accurate spatial and temporal coherences with excellent stability in terms of the output power.

Wavefront-splitting interferometers

In contrast to the amplitude-splitting interferometers, grating fabrication based on the wavefront (sometimes called front wave)-splitting interferometers is not as popular. However, these specific interferometers have number of effective advantages over the amplitude splitting ones. Among others, the prism interferometer [32,33] and the Lloyd's interferometer [34] are two wavefront dividing interferometers that have been used to write the Bragg gratings in optical fibres.

Figure 5.5 shows a detailed schematic of the prism interferometer [32,34] employed in building Bragg grating. The prism is formed from high-homogeneity UV grade-fused silica ensuring high quality of characteristics transmission. In this arrangement, the UV light beam is extended by refraction laterally at the input face of the prism. The extended beam is spatially divided in two by the prism edge. One beam (half of the initial one) is spatially inverted by total internal reflection from the prism face. At the output face of the prism, the two (half) beams are then recombined forming a fringe pattern parallel to the photosensitive fibre core. The lens of the cylindrical shape is put just before the system, helping in the formation of the interference pattern on a line along the optical fibre core.

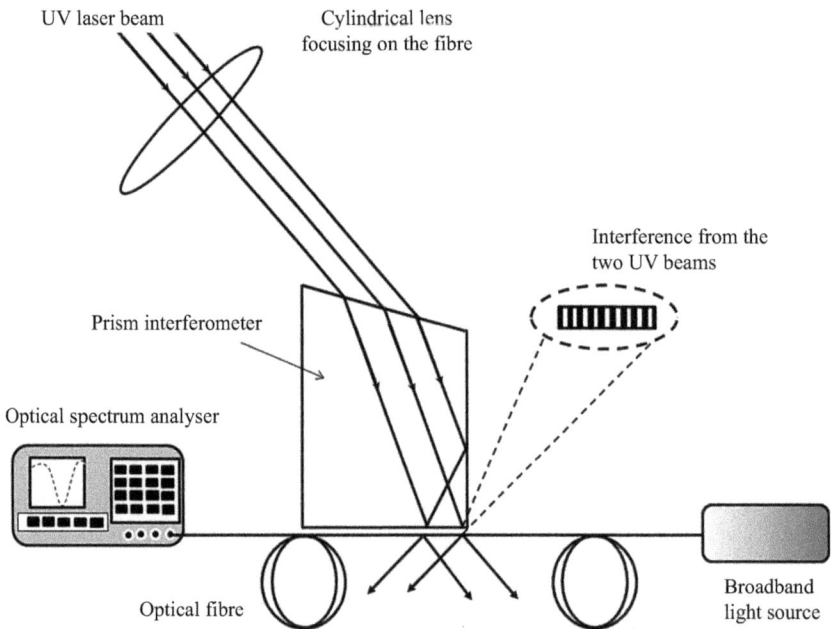

Figure 5.5   Schematic of the prism interferometer employed for the fabrication of Bragg gratings network

Figure 5.6 shows a schematic of the experimental set-up used for the fabrication of Bragg gratings network by means of the Lloyd interferometer. The main constituents of this interferometer are the dielectric mirror that is redirecting 50% of the UV laser beam to an optical fibre that is located perpendicular to it. This UV laser is hence centred at the crossing of the mirror surface and the fibre. On the other hand, the overlapping between the direct part and the redirected portion of the UV light generates interference fringes that are normal (i.e. at 90°) to the fibre axis. Like the precedent interferometers, lens with a cylindrical shape is commonly implanted at face of the system to focus the fringe pattern along the core of the optical fibre.

The main advantage of this wavefront-splitting interferometer is the fact that only one optical element is employed. This considerably reduces the degrees of sensitivity to mechanical vibrations. Moreover, the short distance where the UV laser beams are split decreases the wavefront distortion that is generally induced by dry air electrostatic currents and temperature gradient between the two interfering beams. Additionally, for the wavelength tuning, this assemblage can be easily rotated to change the angle of the beams intersection. The main limitation of this system is the fact that the grating length is restricted to 50% of the beam width. Additional disadvantage is the restricted range of Bragg wavelength tunability, dictated by the physical arrangement of the interferometer itself. In fact, as the

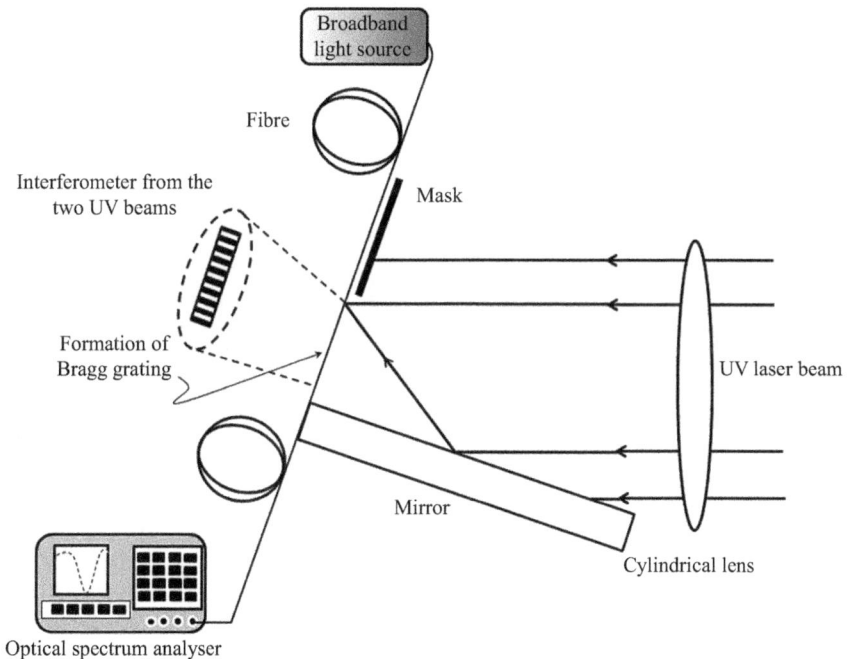

Figure 5.6   Schematic of the Lloyd interferometer used for the fabrication of the Bragg gratings network

crossing angle increases, the difference between the beam path lengths increases, consequently, the beam coherence length restricts the Bragg wavelength tunability. Figure 5.7 shows an experimental set-up of an UV laser excimer pump dye (245 nm wavelength) with a frequency doubled beta-barium borate (BBO) crystal employed for inscribing Bragg gratings.

5.4.1.2   Laser source requirements

Optical laser sources commonly used for writing Bragg gratings through the above described interferometric techniques need to show an excellent spatial and coherences. The spatial coherence conditions can be tolerated in the scenario when an amplitude-split interferometer is used, by simply ensuring that the total number of light reflections is the same in both arms. This element is particularly crucial when we employed a laser beam with a low spatial coherence, as for an UV excimer laser. On the other hand, the temporal coherence has to cover the length of the grating to offer a high contrast ratio for the interfering beams, which in turn could result in enhanced quality of the Bragg gratings. This particular requirement for the coherency and the 240–250 nm needed UV wavelength range force researchers to use complicated laser systems.

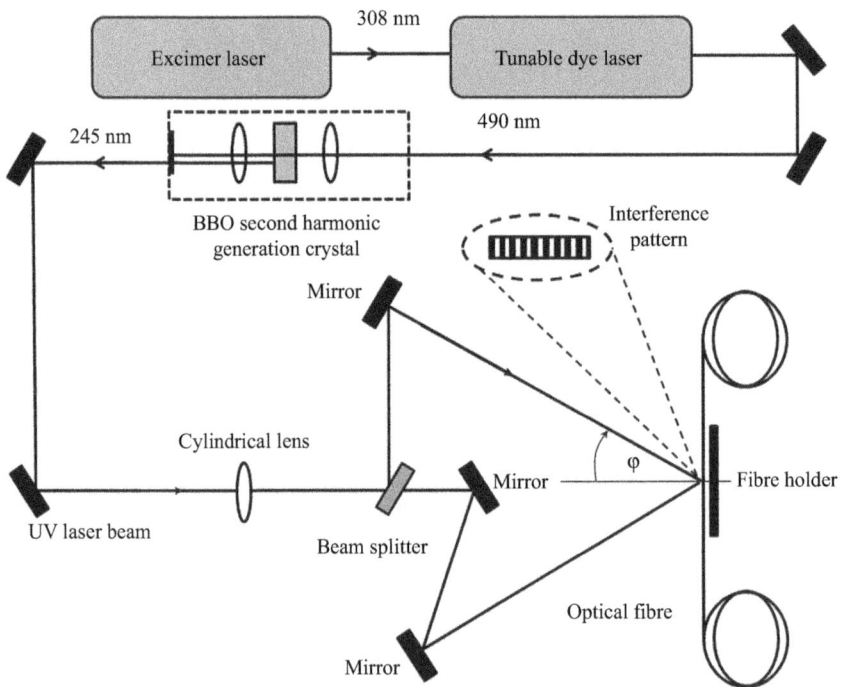

Figure 5.7   Experimental set-up of an UV excimer pump dye laser, wavelength 245 nm, with a frequency-doubled BBO crystal used for the writing of Bragg gratings network in an interferometer

This operating tool consists of a tunable UV excimer dye laser, with a spectral range between 480 and 500 nm wavelengths. The dye laser is then irradiated on a non-linear crystal to double the frequency (Figure 5.7). We provide through this set-up 10–20 ns pulses (based on the laser pump), and about 3–5 mJ with a high spatial and temporal coherences. Another option to this complicated arrangement is a particularly arranged excimer laser with a long-temporal coherence length.

These particular excimer lasers which are offering such narrow spectral line width could operate for comprehensive periods of time with minimum changes in their optical characteristics using the same gas mixture. The commercial available excimer laser systems offering a narrow line-width are usually fabricated based on oscillator amplifier, which rend them highly costly. Othonos *et al.* [35] have successfully developed a simple and cost-effective technique where KrF excimer lasers are modified with a spectral narrowing system for writing Bragg gratings in a side of a subscribed interferometric configuration. A commercial KrF excimer laser (Lumonics Ex-600) was refitted to generate a spectrally narrow beam (Figure 5.8) showing a line width of about 4×10^{-9} mm. This modified laser was then employed to successfully write in photosensitive optical FBGs [35].

An alternative to this described system is what is called the interactivity frequency-double Ar ion laser [36], which is becoming very popular. This laser uses BBO and efficiently transforms high-power visible wavelengths into deep UV laser (244 and 248 nm). The typical features of these lasers include narrow line width, excellent beam stability and unique spatial coherence; all these characteristics rend these systems very attractive for inscribing Bragg gratings in optical fibres [36].

5.4.2 Phase-mask technique

The phase-mask process was reported as one of the most effective techniques for fabricating Bragg gratings in photosensitive optical fibre [37,38]. This technology

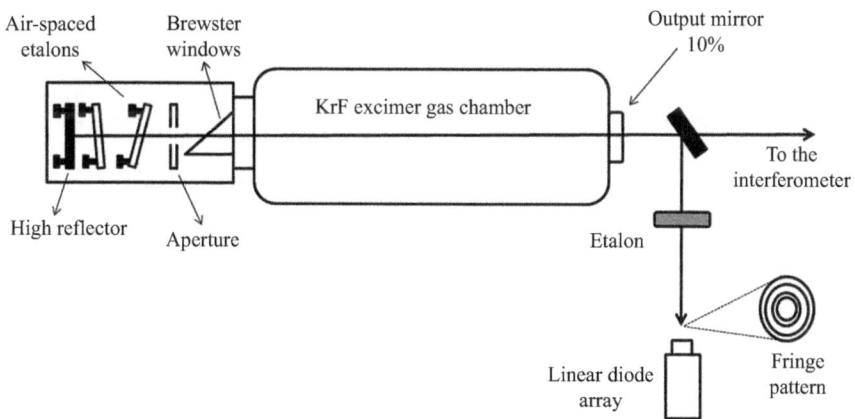

Figure 5.8 Schematic of the experimental set-up used for the arrangement and/or for narrowing the linewidth of an excimer laser beam

uses a phase mask that acts as a diffractive optical element to modulate the UV writing light in the space dimension (Figure 5.9). Electron-beam lithography and holographic processes are the two main techniques forming the phase masks [38]. Holographic-based technology creates phase masks with no stitch error, while it is usually present (i.e. stitch error) in the electron-beam technology [39]. However, note that advanced and complicated patterns are possibly inscribed through the electron beams-based masks, including Moire patterns, quadratic chirps and so on. The phase-mask technology fabricating Bragg gratings has a structure based on one-dimension surface relief that is manufactured in a high quality-fused silica flat which is optically transparent to the UV writing beam. When the laser beam impacts the phase mask, the zero-order diffracted beam is typically cut-off down to few per cent of the transmitted power (below 5%) to select the profile of the periodic gratings. Moreover, the first orders (i.e. the diffracted plus and minus ones) are typically amplified in such a way to contain more than 35% of the transmitted power. The interference of the plus and minus first-order diffracted beams serves also to produce the near-field fringe pattern having a period of equal to one-half of the mask period. The interference pattern footprints an RI modulation in the core of the photosensitive optical fibre that is located in contact with the phase mask or in close proximity immediately behind it (Figure 5.9). Often, a cylindrical lens is employed to focus the fringe pattern along the fibre core.

The use of the phase-mask simplifies considerably the fibre grating fabrication system and increases the stability and the reproducibility of the Bragg grating since only one optical element is used. In fact, as mentioned, the optical fibre is placed

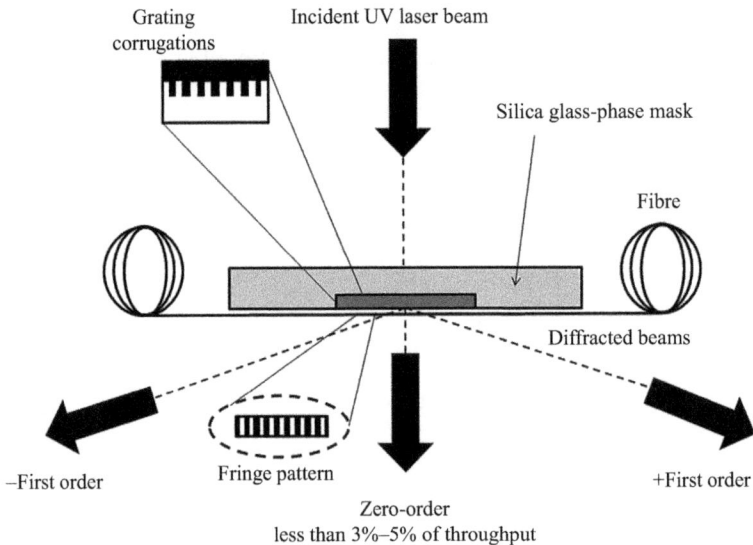

Figure 5.9   Schematic of the phase-mask technique for photo-imprinting an RI Bragg grating in a photosensitive optical fibre

usually directly behind the phase mask in the near field of the diffracting UV light; thus its sensitivity to mechanical vibrations and, therefore, stability problems are considerably minimised. On the other hand, unlike the interferometric technology, here, due to the geometry of the set-up, the writing capability is not impacted by the low temporal coherence. Commercially, the UV KrF excimer lasers sources are the most common to fabricate Bragg gratings using a phase-mask technique. Typically, the UV laser sources show rather a low temporal and spatial coherence which imposes some additional restrictions to the system. As a matter of fact, to induce a maximum modulation in the RI, the low spatial coherence requires placing the optical fibre core in a close contact to the grating corrugations on the phase mask. The modulation of the induced index is proportional to the proximity of the optical fibre from the phase mask, which in turns impacts the Bragg gratings reflectivity. Consequently, the quality of the fabricated gratings depends on the degree of separation between the optical fibre and the phase mask, as a direct contact between them is not advised and may cause serious damages to the phase mask.

Using the phase-mask technique, Othonos and Lee [40] established the significance of the spatial coherence between the UV light beam sources employed to fabricate the Bragg gratings. In fact, elaborating this spatial coherence improves both the quality and the strength of the gratings; it also eases the requirement that the optical fibre may have when it is in direct contact mode with phase mask.

To better appreciate the importance of spatial coherence in the inscription of the Bragg grating through the phase-mask process, it is constructive to regard a simple schematic diagram (Figure 5.10).

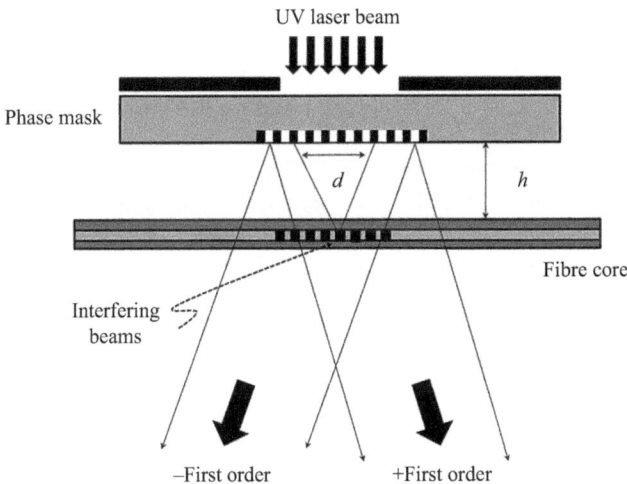

Figure 5.10   Schematic of the phase-mask geometry for the Bragg grating fabrication in the optical fibre core. The plus and minus first-order diffracted beams interferes at the fibre core, placed at a distance h from the mask

Suppose the optical fibre core located at a distance h from the phase mask. The interference of the transmitted plus and minus first orders to create the fringe pattern on the fibre core originates from various parts of the mask (Figure 5.10). As the two interfering UV beams are located at the same distance to the optical, the necessity for temporal coherence is not mandatory for the establishment of a high-contrast fringe pattern. However, as the distance h increases, the separation d between the two interfering beams arising from the phase mask increases as well.

In this scenario, to form a fringe pattern with a high contrast, the necessity for excellent spatial coherence is important. Since the distance h prolongs outside the limits of the spatial coherence of the incident UV laser beam, the contrast of the interference fringe degenerates, resulting ultimately in loosing completely the interference. The crucial need of spatial coherence was also established by Dyer *et al.* [41] who used an excimer UV KrF laser to form gratings in polymeric polyimide film through an irradiated phase mask.

One of the key benefits to locate the optical fibre not against the phase mask is the degree of freedom we could have, e.g., to angle the fibre relative to the mask, producing thereby blazed gratings. In fact, by positioning one extremity of the irradiated fibre section against the phase mask and the second end at a given distance from the mask, this configuration makes changing the induced Bragg grating centre wavelength a possible thing.

From geometrical point of view, one can write a general expression of the centre wavelength tunability of the Bragg grating as

$$\lambda_B = 2n\Lambda\left[\left(1 + \left(\frac{r}{l}\right)^2\right)\right]^{1/2} \tag{5.5}$$

where 'Λ' and l are the fibre grating period and length, respectively, r is the distance between one extremity of the irradiated optical fibre and the phase mask, also called the phase-mask period. Varying a fixed r value in the blazed gratings will vary the centre of the Bragg wavelength.

In an experiment by Othonos and Lee [40], a phase mask with a fibre grating period of 0.531 μm and a length of 100 m were employed, and resulted in λ_B of 1.5580 μm at $r = 0$ (i.e. the optical fibre was directly placed parallel to the phase mask).

Varying the phase-mask pattern with the fibre in close contact to the mask has also been successfully demonstrated [42]. This approach employs a UV transmitting silica prism. The -1 and $+1$ orders are reflected inside a rectangular prism as schematised in Figure 5.11. This non-contact technology is flexible and offers an additional degree of freedom by permitting fast variation of the written Bragg wavelength.

5.4.3   Point-by-point fabrication process of the Bragg gratings

The point-by-point process [43] for inscribing Bragg gratings is achieved by causing a variation in the RI along the core of the optical fibre. Every grating plane is created individually by a focused single-excimer laser pulse, which passes through a mask containing a slit. A focusing lens pictures the slit onto the optical fibre core from the side, as displayed in Figure 5.12, and the RI growth locally in the irradiated fibre section. The optical fibre is then pulled over a distance L, which corresponds to the grating pitch in a way parallel to the fibre axis, and the method is

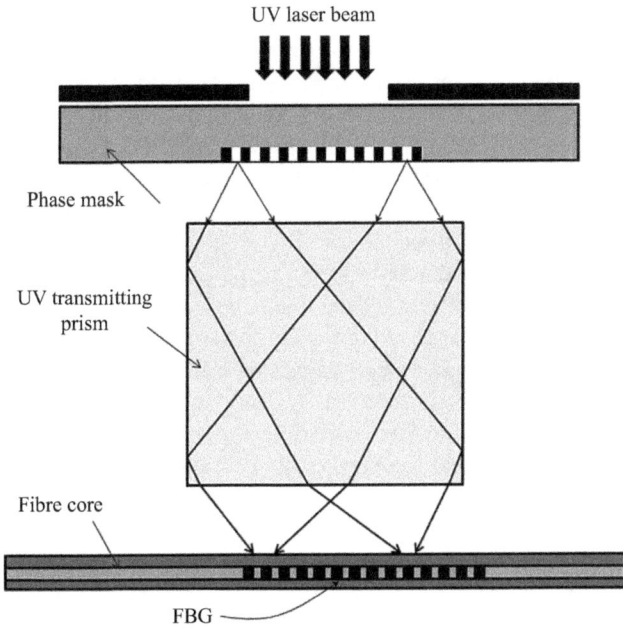

Figure 5.11   Schematic of a non-contact interferometric phase-mask technique for generating fibre Bragg gratings (FBGs)

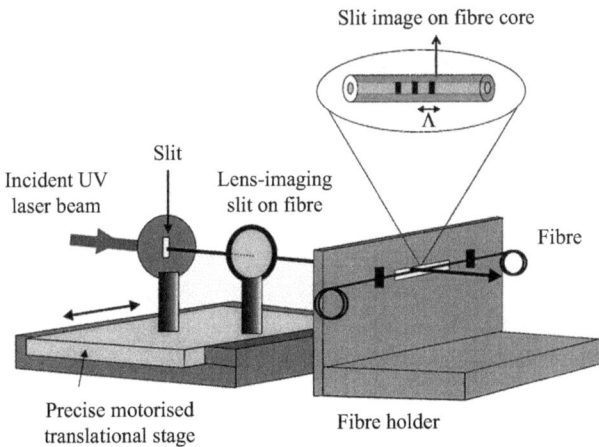

Figure 5.12   Set-up for point-by-point grating fabrication

replicated to write the grating in the fibre core. This point-by-point writing process is also characterised by its high stability and unprecedented accuracy even at the submicron scale.

One of its major advantages is also its flexibility to adjust and control the Bragg grating specifications. As the structure of the Bragg grating is fabricated one point at a time, changes in the grating parameters, including pitch, length and spectral response, can be efficiently implemented. Each time when the fibre is irradiated, the chirped gratings could be easily fabricated with a high precision by expanding the number of fibre translation. The point-by-point model permits the inscription of many modes, including the spatial-mode and the polarisation-mode converters and the rocking filters [44,45], with grating pitch L varying from micrometres to millimetres scale. As the energy of the UV light pulse can be assorted between the points of the induced RI variation, the latter can be controlled in such a way to meet any requested spectral response. Moreover, as this point-by-point fabrication process is a step-by-step method, it necessitates somehow a long processing time, and misstep related to the grating spacing in the fibre's strain that are proper to the thermal and/or small variations effects can appear. This restricts the gratings to a rather very short length. Usually, at 1,550 nm, the period of the grating that is necessary for the first-order reflection is of 530 nm. However, due to the submicron translation and the constraints of the appropriate focusing that is needed, the first-order Bragg gratings at 1,550 nm have not yet been experimentally validated by means of the point-by-point fabrication process. Malo *et al.* [43] successfully inscribe a specific Bragg grating with the ability to reflect the optical signal only through its second and third order. These orders have a grating pitch of about 1 and 1.5 mm, respectively. Figure 5.13 displays the third-order Bragg grating reflection spectrum fabricated by the point-by-point written process.

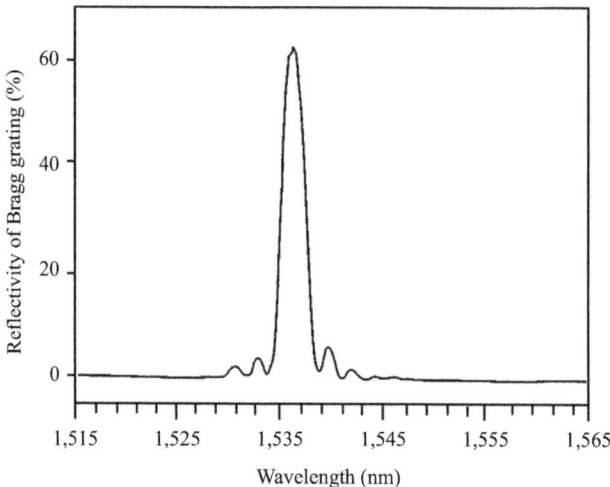

Figure 5.13   Reflection spectrum of a third-order Bragg grating fabricated using the point-by-point method. Adapted, with permission, from [43]

This third-order grating is created with 225 index perturbations, having a grating period $L = 1.59$ mm and hence resulting in a grating length of 357.75 mm. The grating shows a peak reflectivity of 70% located at 1,536 nm and a full width at half maximum of 2.7 nm.

5.4.4 Mask image projection

Further to the famous processes for fabricating FBG reviewed above, high-resolution mask projection approach for fabricating Bragg gratings in optical fibre. through UV excimer-pulsed laser has been also demonstrated [46]. The mass projection structure includes a source of an UV excimer laser beam that is impacting the transmission mask.

On the other hand, the group of Mihailov [46] has used a transmission mask made with a succession of spaced UV opaque lines. The transmitted light was imaged onto the optical fibre core through high resolution-fused silica arrangement. Doing so, using a mask-imaging technique, gratings with various periods ranging from 1 to 6 mm have been successfully inscribed in single-mode Ge-doped fibre. As the set-up and the laser source are simple, the recording of coarse period gratings by this mask-imaging exposure technique may be more flexible than other processes. Blazed, chirped and other complexes grating structures can be easily realised with this approach by a simple change of the used mask.

5.5 Types of fibre Bragg gratings

There are various specific structures of optical FBG including the 'common Bragg reflector', the 'chirped Bragg grating' and the 'blazed Bragg grating'. These Bragg gratings are characterised either by their spacing between the grating planes, known also as grating pitch, or by the inclination angle between these planes and the fibre axis.

The most typical Bragg grating is the Bragg reflector, characterised by its constant pitch. The blazed grating is distinguished by that fact that its phase front is inclined as a function to the fibre axis, in such a way that the angle between the grating planes and the fibre axis is below 90°. The chirped grating is typified by an aperiodic pitch of a monotonic enhancement of in the spacing between grating planes.

A brief overview on these Bragg gratings and their applications is presented in the following sections.

5.5.1 Common Bragg reflector

The common Bragg reflector is the first intracore fibre grating written by the 'self-induced' fabrication technique. It is a simple and the most used FBG and is schematised in Figure 5.14.

Based on various factors, including the length of the Bragg grating and the magnitude of the variation in the induced index, the Bragg reflector can be employed as broadband mirror, a narrowband transmission and/or reflection filter. When combined with another Bragg reflector, the obtained devices can be configured to behave as a bandpass filter. These configurations are presented in Figure 5.15. In this

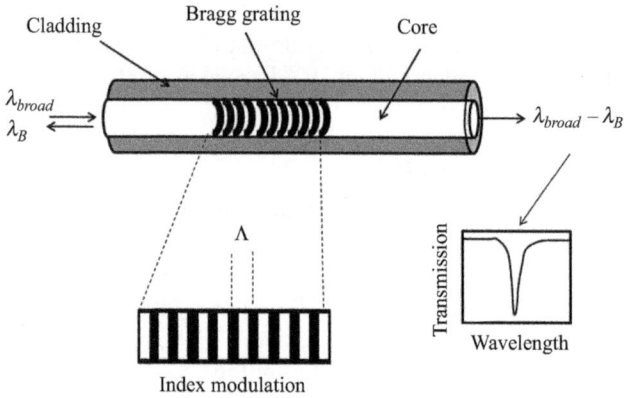

Figure 5.14   Schematic illustrating a uniform Bragg grating with constant index-modulation amplitude and period

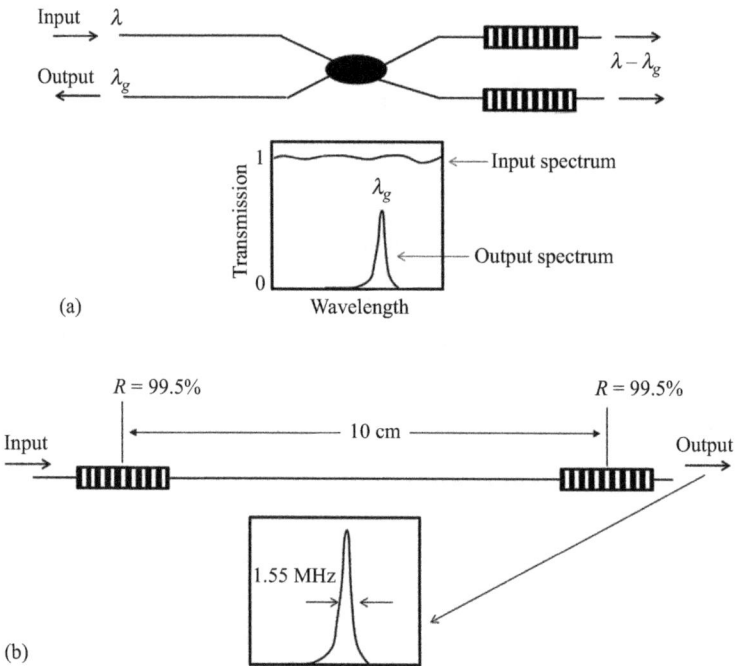

Figure 5.15   Optical fibre bandpass filters using Bragg reflectors: (a) filter arranged in a Michelson-type configuration and (b) filter arranged in a FP-type configuration

type of reflector, the measurements are wavelength encoded, that is why the Bragg reflectors are advised to be excellent temperature and strain sensors.

This fixes the issues related to the fluctuations of both the intensity and the amplitude that characterise the other types of FOS. As each Bragg reflector is assigned with its specific wavelength-encoded signature, a series of the Bragg reflector grating can be inscribed on the same fibre, where each reflector has a specific Bragg resonance wavelength.

This arrangement could be employed as wavelength division multiplexing or quasi-distributed sensors [47].

These Bragg reflectors gratings have also been established to be very valuable components in different lasers as tunable fibre or semiconductor lasers [48–50], by serving as one or even two ends of the laser cavity, depending on the laser system and configuration. They can tune also the laser wavelength by simply changing the Bragg resonance feedback signal.

Ball and Morey [51] successfully demonstrated a single-mode erbium fibre laser that is continuously tunable, where two Bragg grating networks were used as reflectors into a FP arrangement.

This particular continuous tenability property was established without the hopping mode when the enclosed fibre and gratings are stretched uniformly.

FBG lasers can be as well employed as sensors for which the Bragg reflector plays simultaneously both roles of tuning element and sensor [52]. A series of Bragg reflectors with separate and distinguished wavelength-encoded signatures are easily multiplexed in the fibre laser sensor configuration for multi-sensing point [53,54].

5.5.2 Blazed Bragg gratings

Once blazing or tilting (at angles) the Bragg grating planes of the optical fibre axis (Figure 5.16), the transmitted light will be coupled out of the fibre core into a loosely bound guided cladding modes or even into radiation modes outside the fibre. The tilt angle of the grating planes and the amplitude of RI variation determine the efficiency of the coupling and bandwidth of the optical signal that is tapped out.

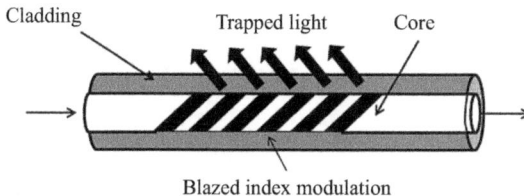

Figure 5.16   Schematic diagram illustrating a blazed grating. Light is directed either upward or downward depending on the propagation direction of the bound mode

The principle to meet the Bragg condition in the blazed grating configuration is the same as that of the Bragg reflector. Figure 5.17 displays the vector diagram of the Bragg condition (i.e. energy and momentum conservation) for the blazed grating. Here, ψ is the angle of incidence of the wave vector of the grating as a function to the fibre axis. From a geometrical point of view, the simple trigonometry shows that the magnitudes of the incident and scattered wave vectors, ξ_i and ξ_s, respectively, must be equal ($\xi = \xi_i = \xi_s$). Hence, the scattered wave vector must be at angle $\xi = 2\psi$ relatively to the fibre axis. The application of the law of cosines to the momentum diagram gives

$$\xi_i^2 + \xi_s^2 - 2\xi_i\xi_s \cos(\pi - 2\psi) = K^2 \tag{5.6}$$

K is the wavevector defined in Figure 5.17, $K = \xi_s - \xi_i$.

Which simplifies to $\cos(\psi) = K/2\xi$ and demonstrates that the scattering angle is governed and limited by both the effective RI and the Bragg wavelength.

Figure 5.17 shows also that for this type of blazed gratings, different wavelengths could rise at different angles, moreover, due to their different propagation constants, different modes of the same wavelength could come out as well at slightly different angles.

Kashyap *et al.* [55] successfully applied the multiple-blazed gratings to level out the gain spectrum of a saturated amplifier based on erbium-doped fibre, and a gain fluctuation of about ±1.6 dB was decreased by more than 81% down to ±0.3 dB. This factor is crucial for fibre-based communications, especially when they employ different optical signals at various wavelengths.

An additional attractive utility of the blazed grating is in 'mode conversion' application, which is manufactured by introducing a periodic physical perturbation in the RI along the fibre length. The periodicity of the RI perturbation links the momentum mismatch between the selected modes and realises a phase-matched coupling between them. Note that various grating periods are employed for mode conversion at various wavelengths.

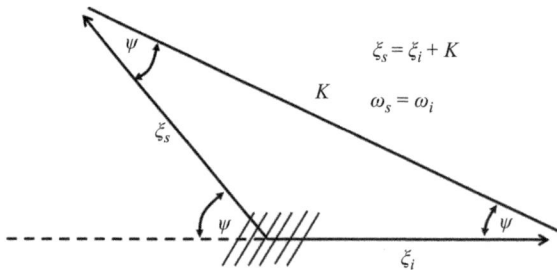

Figure 5.17   Schematic of the vector diagram for the Bragg condition of a blazed grating. The magnitude of the incident ξ_i and scattered ξ_s wave vector are identical

5.5.3 Chirped Bragg grating

This particular type of Bragg grating has a uniform variation of the grating period as depicted in the schematic of Figure 5.18. This variation can be achieved by varying the axis the grating period 'Λ', or by varying the value of the RI, or by variation both of these parameters. The chirped Bragg gratings have been inscribed in optical fibres through different processes [37,56–59]. One of the most known methods is the double exposure model developed by Hill *et al.* [37], succeeding to form a 1.5 cm long chirped grating.

The effective-mode index of the optical fibre was adjusted linearly over the whole grating length by a UV laser radiation (excimer) followed by irradiating the same length a second time through a phase mask to fabricate a linearly chirped grating. A chirped grating of 0.4 nm was successfully demonstrated at 1,549 nm wavelength, with an induced delay of 120 ps over the total bandwidth of the grating period.

A particular simple process based on the phase mask for which the linear chirp is approached by a step chirp was also developed and has become highly quotable to fabricate accurate chirped grating. In this process, cascade of multiple gratings having progressive periods are utilised to mimic a long chirped grating as depicted in Figure 5.18(b). This grating structure is first written on the phase mask. Then, the mask is utilised to inscribe the chirped Bragg gratings into the optical photosensitive fibre. This is definitely a highly controllable process and repeatable approach for fabricating any type of chirped Bragg structure in optical fibre.

For transmission over long distance, as is the case for the long-haul high-bit rate communication systems, the chromatic dispersion induces what is called 'the pulse broadening', which is mainly characterising the erbium-doped amplifier-based optical fibre communication. However, the elimination of this pulse-broadening phenomenon is possible through the use of an element showing chromatic dispersion

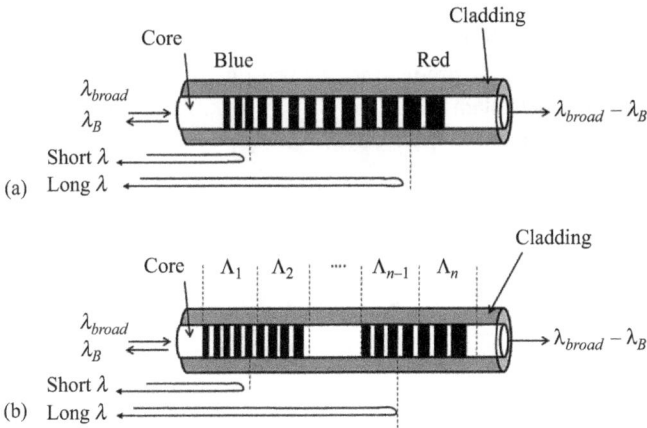

Figure 5.18   Schematic diagram illustrating (a) a chirped grating with an aperiodic pitch and (b) a cascade of several gratings with increasing period

of the same magnitude and opposite sign to that of the optical fibre. As a matter of fact, as the chirped grating is characterised by a resonant frequency which is proportional to the axial location along the grating, and hence different frequencies contained in the light pulse are reflected at different points and, require diverse delay times, it could be therefore utilised as dispersion compensators to compress temporally the broadened pulses (Figure 5.18).

5.5.4   Type II Bragg gratings

Archambault *et al.* [60] conducted experiments based on a pulsed UV excimer laser to study the correlation between the pulse energy and grating strength, L, through a series of single-pulse gratings. The UV laser was focused on a fibre area of about 15×0.3 mm. For each grating, the peak-to-peak index modulation was approximated from the reflection spectrum. It was found the existence of a clear and sharp energy threshold manifested starting from energy of 30 mJ of the laser pulse. Upon this threshold, the photoinduced RI modulation increases drastically. For example, duplifying the energy of the laser pulse from 20 to 40 mJ will result in increasing the photoinduced index modulation by more than thousand times.

Below the threshold level, the RI modulation varies proportionally linearly with the energy density, while above this level, it was found to saturate.

Investigation of a Type II grating with a conventional optical microscopy technique showed a defective track at the interface between the core and the cladding. This destruction track emerges solely in the gratings of Type II, which stipulates that it could be the source behind the large index variation. Moreover, the fact that this damage is restricted only on one side of the fibre core proposes that the major part of the UV optical signal was absorbed, highly likely never reaching the other core side. Even though the exact basis of this process is not yet fully understood, Russell *et al.* [9] suggested that the phenomenon is triggered by the high single-photon absorption occurring at 248 nm, thereby exciting electrons into the conduction band of silica, giving rise to free electrons plasma.

This will cause then the production of a brusque rise in the UV light absorption and occurs a permanent damage in the glass matrix. A schematic detailing of the Type II grating is depicted in Figure 5.19.

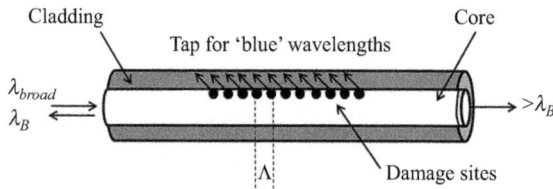

Figure 5.19   Type II grating: Schematic diagram highlighting the deterioration track on one side of the fibre core. The wavelengths that are longer than the Bragg centre wavelength are hence transmitted, while the shorter ones are strongly coupled into the cladding

Bragg gratings of Type II show an extremely high reflectivity and a large bandwidth. Moreover, their transmission of the wavelengths is longer than that of the Bragg centre wavelengths. They strongly couple shorter wavelengths into the fibre cladding, which allows the gratings to operate as an effective selective wavelength valves. The tests performed on the long-term stability have revealed that Type II could be exceptionally stable especially at high temperatures. As a matter of fact, at 800°C, no degradation of the grating reflectivity property was noticed, even after a period of 24 h. The reflectivity property of the grating decreases slowly at 900°C, until a permanent component emerges. At 1,000°C, the majority of the grating vanishes after only 4 h. This brings proof that the mechanism responsible for the high reflectivity Type II single-pulse gratings contrast from the conventional-type model. The stability at high temperature of the Type II gratings rends them beneficial for sensing applications, especially in harsh environment conditions. One of the most interesting characteristics of Type II gratings is the very short time of the highly reflective gratings formation which could be achieved during a single pulse of excimer laser (few nanoseconds). This is of huge practical relevance for the large production of strong gratings during the fibre inscription process, before the application of the protective polymer cladding. Even though the principle of single-pulse Type I and Type II Bragg gratings fabrication during the fibre inscription process has been successfully established [61,62], the quality of the in-line gratings still requires amelioration. One obvious advantage of producing FBGs during the inscription process is that in-line fabrication avoids a potential possible direct contact with the pristine outer surface of the glass, while the offline fabrication requires to strip off the polymer coating section of the fibre in order to be able to write the grating. Because of the surface contamination, the fibre is extremely weakened at the grating location, even if it was subsequently recoated.

5.5.5   Novel structures of the Bragg grating

5.5.5.1   Superimposed multiple Bragg gratings

Othonos *et al.* [63] recently established the possibility of the fabrication of various Bragg gratings on an optical fibre at the same location.

This is of primary importance in fabricating advanced devices, including sensors systems, fibre communications and lasers. Due to the fact that various Bragg gratings at the same location basically accomplish a comb-like function, these devices are ideally appropriate for multiplexing and demultiplexing functions, requiring a minimum of space, offering an additional advantage to the optical-integrated technology. This is truer when the size is often an issue. This system could also be employed for the detection of material where the multiple Bragg lines are designed in such a way to match the spectral signature of a defined material. Additional appealing observation is that the CW of the existing Bragg gratings was red-shifted to longer wavelengths each time a new grating was formed due to the effective RI variation.

5.5.5.2   Superstructure Bragg gratings

This type of Bragg grating structure is attributed to a grating manufactured with a modulated light irradiation over an elongated length on the gratings [64]. Eggleton *et al.* [64] employed this process to transform the UV inscription beam along the fibre and phase-mask assembly during the modulated of the laser beam (an excimer dye laser of 2.0 mJ delivered at 240 nm). Hydrogenated boron co-doped single-mode optical fibre was first placed in contact with the phase mask, then the UV laser irradiating a 1 mm length was focused *via* the phase mask into the fibre core. The laser beam was periodically triggered at a repetition rate of 10 Hz frequency for 15 s 'ON' to produce bursts of 150 pulses and translated at a determined velocity of 0.19 mm/s along the mask, permitting thereby to write a 40 mm-long superstructure.

5.5.5.3   Phase-shifted Bragg gratings

The Bragg gratings in the optical fibre operate habitually as reflection filters having a narrowband which is centred at the Bragg wavelength due to the stop-band that is affiliated to the one-dimensional periodic medium. Numerous applications, including the channel selection in a communication multichannel scheme would improve significantly if the Bragg gratings could be constructed as a transmission filter having a narrowband. Despite the fact that many technologies using the model of Michelson and FP interferometers have been successfully deployed for this aim [65], their functionality still requires multiple Bragg gratings and could induce additional losses. However, a model usually employed in distributed feedback (DFB) semiconductor lasers [66,67] could be also employed to design the transmission spectrum that meets specific needs. The technology reposes on introducing a phase shift through the optical fibre grating with a magnitude and location that are adaptable to intend a particular transmission spectrum. In fact, this model was inspired from the work of Haus and Shank [68] first proposed in 1976. The experimental precept of the phase shift has been established by Alferness *et al.* [69] during the 1980s (1985–86) using periodic structures fabricated from semiconductors. The phase shift was then fabricated *via* the etching of a larger spacing at the centre of the fibre, forming thereby the basis of what is known today as single-mode phase-shifted semiconductor lasers [70]. An analogous device may be assembled in the optical fibres through the following different processes:

1. Phase-masks technique, in which the locations of the phase shift was inscribed into the mask design [71].
2. Irradiation of the grating region by a UV-pulsed laser beam (i.e. post-processing of the grating) [72].
3. Post-fabrication processing through a localised heat-treatment technique [72]. Such treatment produces two gratings which are out of phase with each other and operates as a wavelength selective FP resonator.

The resonance wavelength is a function of the size of the phase variation [73]. One of the most straightforward applications includes the fabrication filters with an ultra-narrowband transmission and reflection. In addition, shift of multiple phases can be processed and integrated to produce particular devices such as comb filters,

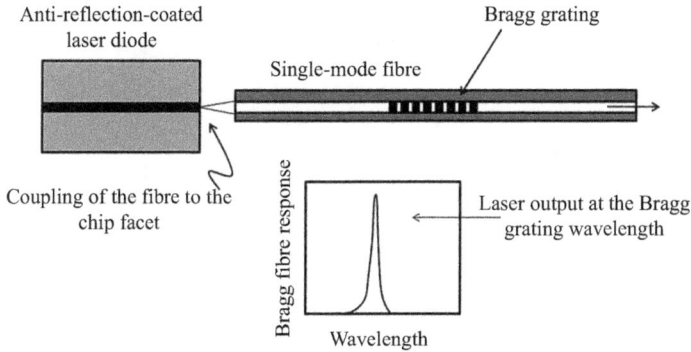

Figure 5.20   *Schematic of the external fibre grating-based semiconductor laser: the semiconductor laser chip is coated by an anti-reflection layer on the output face and coupled to the optical fibre with a Bragg grating, forcing hence the oscillation at the Bragg grating wavelength*

or employed to ensure the single-mode operation of DFB fibre lasers. Figure 5.20 is a schematic of a semiconductor laser based on an external fibre grating that is coated by an anti-reflection polymer on the output face.

5.6   Fibre Bragg grating encapsulation

Due to the mechanical fragility of FBG bare sensors, it is somehow difficult to employ them directly in any infrastructure without a protection. Hence it is inevitable to develop an encapsulation process for bare FBG strain sensors.

Various types of encapsulation technologies were recently developed for bare FBG strain and temperatures sensors. These protection coatings were found to be stable for temperatures as high as 1,000°C when, e.g., using gold-coated fibres with special packaging that is ideally balancing between the protection capability for the passive fibre and the fast thermal conductivity on the sensor locations.

To choose the ideal packaging design for the FBG sensors, a series of optical fibres were fabricated and systematically tested in various small tubes of less than 1 mm outer diameter. The best option was to put only one optical fibre within small thin stainless steel tube of 0.35/0.45 mm input/output diameters. This design allowed a fast transient and a strong enough thin wall to prevent the tube buckling. Examples of packaged-coated fibre sensor are depicted in Figures 5.21–5.23.

The capillary encapsulation of the FBG-based strain sensor can be favourably applied in many civil engineering structures including concrete ones. The holder ring is employed to maintain the deformation of the FBG sensor proportional to the

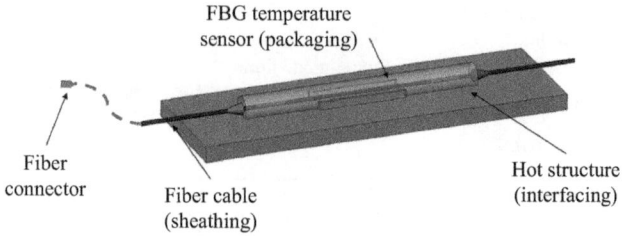

*Figure 5.21   Sketch of the capillary encapsulation of FBG-based temperature
sensor*

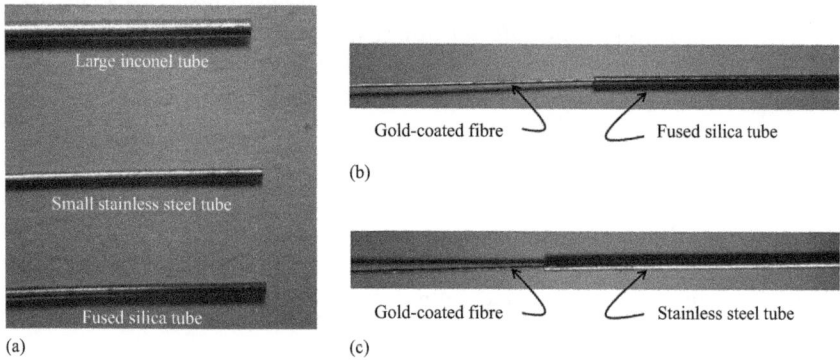

*Figure 5.22   Typical optical photos of various packaged FBG including
(a) Inconel, stainless steel-fused silica tubes; (b) gold-coated
FBG encapsulated by fused silica tube and (c) gold-coated
FBG encapsulated by stainless steel tube (photo courtesy of
MPB Communications, Inc.)*

concrete structures one, and the elongated optical fibre is available for the temperature compensation connector.

FBG-based strain sensors are also encircled in slice protected by commercial glue, which can be easily employed efficiently on the top of the metal surface and concrete structures.

It is noteworthy to mention that the encapsulated FBG sensors should be calibrated before the sensors are applied in practical structures, because often this encapsulated layer will exert a pressure and alter the sensitivity of the original FBG.

Finally, fibre-reinforced polymers (FRP) are presently largely accepted for many advanced applications as a new type of valuable construction material. As a matter of fact, to take full advantages both from employing the FRP (i.e. in terms of using their strength properties) and FBG-sensing properties, Chan *et al.* [74] reported the manufacturing process of FRP optical FBG bars. The FRP bars were

(a)

| Fused silica tube | Stainless steel large tube | Stainless steel small tube | Gold-coated fibre |

(b)   Details (microscope ×25)   Details (microscope ×25)

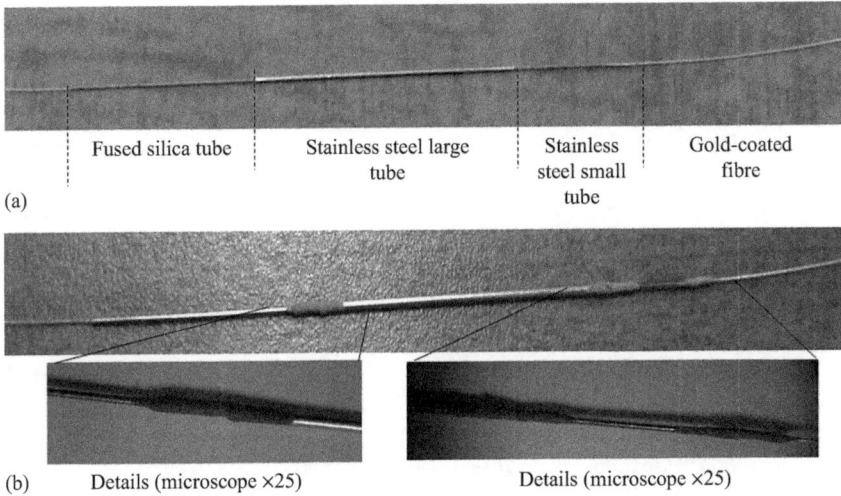

Figure 5.23   Typical optical photo of packaged (a) gold-coated FBG-encapsulated stainless steel (small and large tubes) and (b) gold-coated FBG encapsulated epoxy bonding for high-temperature sensing (photos courtesy of MPB Communications, Inc.)

simply impacted with optical FBG and were shown that this does not alter the mechanical properties as the FBG diameter is much smaller than the FRP bars one.

References

[1]   K.O. Hill, Y. Fujii, D.C. Johnson and B.S. Kawasaki, *Applied Physics Letters*, 1978, **32**, 647.

[2]   B.S. Kawasaki, K.O. Hill, D.C. Johnson and Y. Fujii, *Optics Letters*, 1978, **3**, 66.

[3]   G. Meltz, W.W. Morey and W.H. Glenn, *Optics Letters*, 1989, **14**, 823.

[4]   K.O. Hill, B. Malo, F. Bilodeau and D.C. Johnson, *Annual Review of Materials Science*, 1993, **23**, 125.

[5]   R.J. Campbell and R. Kashyap, *International Journal of Optoelectronics*, 1994, **9**, 33.

[6]   R. Kashyap, *Optical Fibre Technology*, 1994, **1**, 17.

[7]   D.K.W. Lam and B.K. Garside, *Applied Optics*, 1981, **20**, 440.

[8]   J. Stone, *Journal of Applied Physics*, 1987, **62**, 4371.

[9]   P. St. J. Russell, J.-L. Archambault and L. Reekie, *Physics World*, 1993, **6**, 41.

[10] J.M. Yeun, *Applied Optics*, 1982, **21**, 136.

[11] D.L. Williams, S.T. Davey, R. Kashyap, J.R. Armitage and B.J. Ainslie, *Electronics Letters*, 1992, **28**, 369.

[12] R.M. Atkins and V. Mizrahi, *Electronics Letters*, 1992, **28**, 1743.

[13]   D.P. Hand and P.S. Russell, *Optics Letters*, 1990, **15**, 2, 102.

[14]   K.O. Hill, B. Malo, F. Bilodeau, *et al.*, *Proceedings of the Conference on Optical Fibre Communications*, 1991, **14**, PD3-1.

[15]   M.M. Broer, R.L. Cone and J.R. Simpson, *Optics Letters*, 1991, **16**, 1391.

[16]   F. Bilodeau, D.C. Johnson, B. Malo, K.A. Vineberg and K.O. Hill, *Optics Letters*, 1990, **15**, 1138.

[17]   D.L. Williams, S.T Davey, R. Kashyap, J.R. Armitage and B.J. Ainslie, *Electronics Letters*, 1992, **28**, 369.

[18]   P.T. Taunay, P. Niay, P. Bernage, *et al.*, *Optics Letters*, 1994, **19**, 17, 1269.

[19]   P.J. Kaiser, *Journal of the Optical Society of America*, 1974, **64**, 475.

[20]   E.J. Friebele, C.G. Askin, M.E. Gingerich and K.J. Long, *Nuclear Instruments and Methods in Physics Research Section B Beam Interactions with Materials and Atoms*, 1984, **1**, 355.

[21]   H. Kawazoe, Y. Watanabe, K. Shibuya and K. Muta, *Materials Research Society Symposium Proceedings*, 1986, **61**, 350.

[22]   P. St. J. Russell, L.J. Poyntz-Wright and D.P. Hand, *Proceedings of SPIE*, 1990, **1373**, 126.

[23]   D.L. Griscom and E.J. Friebele, *Physical Review B*, 1981, **24**, 4896.

[24]   P.J. Lemaire, R.M. Atkins, V. Mizrahi and W.A. Reed, *Electronics Letters*, 1993, **29**, 1191.

[25]   F. Bilodeau, B. Malo, J. Albert, D.C. Johnson and K.O. Hill, *Optics Letters*, 1993, **18**, 953.

[26]   D.L. Williams, B.J. Ainslie, R. Armitage, R. Kashyap and R. Campbell, *Electronics Letters*, 1993, **29**, 45.

[27]   J. Albert, B. Malo, F. Bilodeau, *et al.*, *Optics Letters*, 1994, **19**, 387.

[28]   P.E. Dyer, R.J. Farley, R. Giedl, K.C. Byron and D. Reid, *Electronics Letters*, 1994, **30**, 860.

[29]   M.G. Sceates, G.R. Atkins and S.B. Poole, *Annual Review Materials Sciences*, 1993, **23**, 381.

[30]   D.L. Williams, B.J. Ainslie, R. Kashyap, *et al.*, *Proceedings of SPIE*, 1993, **2044**, 55.

[31]   C. Fiori and R. Devine, *Materials Research Society Symposium Proceedings*, 1986, **61**, 187.

[32]   R. Kashyap, J. Armitage, R. Wyatt, S. Davey and D. Williams, *Electronics Letters*, 1990, **26**, 730.

[33]   B. Eggleton, P. Krug and L. Poladian, *Optics Letters*, 1994, **19**, 877.

[34]   H. Limberger, P. Fonjallaz, P. Lambelet, R.P. Salathe, Ch. Zimmer and H. Gilgen, *Proceedings of SPIE*, 1993, **2044**, 272.

[35]   A. Othonos and X. Lee, *Review of Scientific Instruments*, 1995, **66**, 3112.

[36]   J. Cannon and S. Lee, *Laser Focus World*, 1994, **2**, 50.

[37]   K. Hill, B. Malo, F. Bilodeau, D. Johnson and J. Albert, *Applied Physics Letters*, 1993, **62**, 1035.

[38]   D.Z. Anderson, V. Mizrahi, T. Erdogan and A.E. White, *Electronics Letters*, 1993, **29**, 566.

[39] J. Albert, S. Theriault, F. Bilodeau, D. Johnson, K. Hill, P. Sixt and M. Rooks, *IEEE Photonics Technology Letters*, 1996, **8**, 1334.

[40] A. Othonos and X. Lee, *IEEE Photonics Technology Letters*, 1995, **7**, 1183.

[41] P. Dyer, R. Farley and R. Giedl, *Optics Communications*, 1995, **115**, 327.

[42] R. Kashyap, G.D. Maxwell and B.J. Ainslie, *IEEE Photonics Technology Letters*, 1993, **5**, 2, 191.

[43] B. Malo, K. Hill, F. Bilodeau, D. Johnson and J. Albert, *Electronics Letters*, 1993, **29**, 1668.

[44] K. Hill, B. Malo, K. Vineberg, F. Bilodeau, D. Johnson and I. Skinner, *Electronics Letters*, 1990, **26**, 1270.

[45] K. Hill, F. Bilodeau, B. Malo and D. Johnson, *Electronics Letters*, 1991, **27**, 1548.

[46] S. Mihailov and M. Gower, *Electronics Letters*, 1994, **30**, 707.

[47] W. Morey, J. Dunphy and G. Meltz, *Proceedings of SPIE*, 1991, **1586**, 216.

[48] G. Ball, W. Morey and W. Glenn, *IEEE Photonics Technology Letters*, 1991, **3**, 613.

[49] F. D'Amato and J. Dunphy in *Proceeding of the IEEE LEOS Conference Proceedings*, Paper DLTA 7.3, Boston, MA, USA, 16–19 November 1992.

[50] G. Ball and W. Morey, *IEEE Photonics Technology Letters*, 1991, **3**, 1077.

[51] G. Ball and W. Morey, *Optics Letters*, 1992, **17**, 420.

[52] A. Othonos, A. Alavie, S. Melle, S. Karr and R. Measures, *Optical Engineering*, 1993, **32**, 2841.

[53] A. Kersey and W. Morey, *Electronics Letters*, 1993, **29**, 112.

[54] A. Alavie, S. Karr, A. Othonos and R. Measures, *IEEE Photonics Technology Letters*, 1993, **5**, 1112.

[55] R. Kashyap, R. Wyatt and P. Mckee, *Electronics Letters*, 1993, **29**, 1025.

[56] K. Byron, K. Sugden, T. Bircheno and I. Bennion, *Electronics Letters*, 1993, **29**, 1659.

[57] K. Sugden, I. Bennion, A. Molony and N. Copner, *Electronics Letters*, 1994, **30**, 440.

[58] V. Marquez-Cruz and J. Albert, *Journal of Lightwave Technology*, 2015, **33**, 3363.

[59] R. Kashyap, P. McKee, R. Campbell and D. Williams, *Electronics Letters*, 1994, **30**, 996.

[60] J. Archambault, L. Reekie and P. Russell, *Electronics Letters*, 1993, **29**, 453.

[61] L. Dong, J. Archambault, L. Reekie, P. Russell and D. Payne, *Electronics Letters*, 1993, **29**, 1577.

[62] C. Askins, M. Putnam, G. Williams and E. Friebele, *Optics Letters*, 1994, **19**, 147.

[63] A. Othonos, X. Lee and R. Measures, *Electronics Letters*, 1994, **30**, 1972.

[64] B. Eggleton, P. Krug, L. Poladian and F. Quellette, *Electronics Letters*, 1994, **30**, 1620.

[65] W. Morey, G. Ball and G. Meltz, *Optics & Photonics News*, 1994, **5**, 8.

[66] G. Agrawal and A. Bobeck, *IEEE Journal of Quantum Electronics*, 1988, **24**, 2407.

[67] G. Agrawal and N. Dutta in *Semiconductor Lasers*, Van Nostrand Reinhold, New York, NY, USA, 1993.

[68] H. Haus and C. Shank, *IEEE Journal of Quantum Electronics*, 1976, **QE-12**, 352.

[69] R. Alferness, C. Joyner, M. Divino, M. Martyak and L. Buhl, *Applied Physics Letters*, 1986, **49**, 125.

[70] K. Utaka, S. Akiba, K. Skai and Y. Matsushima, *IEEE Journal of Quantum Electronics*, 1986, **QE-22**, 1042.

[71] R. Kashyap, P. McKee and D. Armes, *Electronics Letters*, 1994, **30**, 1977.

[72] J. Canning and M. Sceats, *Electronics Letters*, 1994, **30**, 1344.

[73] D. Uttamchandani and A. Othonos, *Optics Communications*, 1996, **127**, 200.

[74] Y.W.S. Chan and Z. Zhou, *Pacific Science Review*, 2014, **16**, 1.

Chapter 6

Fibre Bragg grating sensors for micrometeoroids and small orbital debris

Space debris refers to all kinds of non-functional objects left in space by man, while the micrometeoroids are occurring naturally from the fragmentation of asteroids and comets. Globally, they are known by micrometeorites and orbital debris (MMOD). At low Earth orbit altitudes (below 2,000 km), the debris number is much larger than the micrometeoroids and constitute a large continuous risk to satellites, spacecraft and the International Space Station (ISS). The risk is increasing with more satellites are launched every year, and others getting to their end of life. The United Nations (UN) put the debris risk mitigation as one of their main objectives in the Committee on the Peaceful Uses of Outer Space. This chapter addresses the monitoring and mitigation of the effects of small debris between 1 and 10 mm, using fibre optic sensors (FOS). Examples of such effects are:

- holes and craters in the honeycomb(s) (HC) panels, these panels are usually made of aluminium and/or carbon-fibre-reinforced polymer(s) (CFRP);
- holes and craters reducing the solar panels efficiency, due to their large surfaces, the solar panels are very exposed to the debris;
- leak in the pressure vessels/tanks (for storing propellant at low pressure, or storing inert gas at high pressure);
- causing vibrations in the steering/pointing mechanisms leading to errors in the guidance, these effects are important in microsatellites;
- damaging the electrical harness;
- creating malfunctions in electronic instruments;
- depressurisation of manned modules; and
- destruction of windows/viewports.

The parameters that can be observed to identify and characterise the debris and the impact are:

- the impact energy of the debris coming with hypervelocity up to 15–20 km/s,
- the impact time and its duration is of the order of microseconds,
- the impact location and distribution of the debris fragments,
- the impact crater or hole and its size and shape,
- the debris size and composition,
- the propagation of the impact effect in the target [acoustic emission (AE), vibration and temperature], and
- shock wave study (very fast event needing acquisition at a nanosecond level).

6.1   Shield against micrometeorites and orbital debris

Bumper shields and energy-absorbing materials are widely used aircraft and auto-motive applications to minimise the damage caused by impacts, enhance safety and reduce the weight requirements for the protection of manned structures.

In space, the MMOD have a hypervelocity of the order of a few km/s, making it very difficult to reduce their impact effect. The basic shielding called 'Whipple shield' has been proposed by Fred Whipple in 1940 [1]. Its basic configuration consists of a bumper, a gap or spacing and a back wall. The role of the first wall or 'bumper' is to break up the direct projectile into smaller fragments (kind of a cloud of material) containing both projectile and wall (or bumper) debris. This cloud then enlarges during its movement through the standoff, resulting in the impact or momentum becoming distributed over a larger area of the back wall. The rear wall has to be thick enough to resist to the blast weight from the fragments cloud and any solid debris that remain [1]. Variants of the Whipple shield are used to protect specific instruments and components on the ISS; they contain additional middle layers of strong flexible materials such as Kevlar® or Nextel.

The two main drawbacks of the Whipple shields are the required large size/volume and mass. Recent studies (Table 6.1) propose alternative structures and

Table 6.1   Structures used for hypervelocity mitigation

Structures	Details
Whipple	Widely used in space, consisting of multilayer metal structures (with certain distance between two consecutive layers) to protect the aircraft from hypervelocity impact. Its main disadvantage is the distance needed between the layers making the volume large and sometime not practical, in addition to the high metal mass
Sandwich	Compact multilayer combining materials with different properties to better stop the debris (metal/glass/composites) – the objective is to reduce the total volume of a Whipple as well as its mass
Honeycomb	The honeycomb-shaped structure provides a material with minimal density and good mechanical properties a high out-of-plane compression and high out-of-plane shear. It can be used as crash absorber for kinetic energy. Aluminium honeycomb is a light weight, environment-friendly (recyclable) material with low cost
Geometrical forms	Composite pyramidal or lattice formed of strips with specific height, angle and distance between the core. The lattice forms the sandwich structure for anti-impact between two plates. Another approach is a matrix composite formed into the geometrical shape of an egg boxes
Foam	Porous structure with low density and good anti-impact effect, e.g. nickel/aluminium hybrid metal foam and witness plate
Laminate	Made by sticking together two or more layers of a particular composite [2]. The composite can be CFRP, the resin holding the layer together and include CNT and self-healing microcapsules. Kevlar, Nextel and carbon fibre-reinforced silicon carbide were proposed
Hybrid metal composites	CFRP laminates embedding metallic fibres providing stronger resistance to debris impact with lower mass compared with complete sheets

configurations, such as HC sandwiched structures, porous metallic foam and short lattices between two aluminium plates or HC. Figure 6.1 shows the Whipple shield. The three stages fixed at 10 cm between two consecutive stages are top, middle and bottom plates. The first is pierced with a diameter about 2–3 cm, the middle is pierced with a diameter about 4 cm by large fragments, with craters (1–5 mm diameter) caused by small fragments. The third plate completely stopped the fragmented pellet, it has a crater bump of around 3 cm diameter with other small craters (up to 1 cm) caused by small fragments.

The use of composites in space structures has largely spread, it can be noticed through the number of published articles dedicated to study their reliability, health monitoring in space and their response to debris. As a specific objective, we look to increase the lifetime of the composite overwrapped pressure vessel(s) (COPV); they are widely used on various spacecraft (e.g. in Falcon 9, Challenger) and ISS. Figure 6.2 shows photography of a carbon fibre COPV experienced at National Aeronautics and Space Administration (NASA). Due to their unique strength-to-weight ratio, COPV are attractive candidates for various space applications as carriers of high-pressure fluids and propellants. Moreover, strong debris damper (Kevlar, Nextel) and self-healing materials (ethylene methacrylic acid) can be added to their protective layers significantly enhancing their competitive advantages of safety,

Figure 6.1   Pictures of a whipped shield of 35 mm thick Al plates, with 0.5 mm thick composite laminates after being hit by 1.25 cm diameter pellet at 2.5 km/s (photo courtesy of MPB Communications, Inc.)

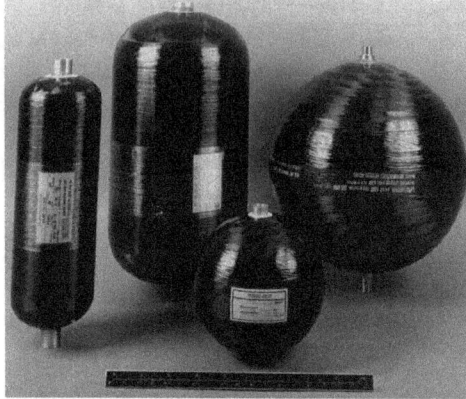

Figure 6.2   Example of carbon fibre COPV experienced at National
Aeronautics and Space Administration (NASA). © 2013, NASA.
Reproduced, with permission, from [3]

reliability lifespan and high-quality performance. For the COPV, the self-healing should also keep the tank hermetic by completely filling the cracks or holes created by the impact and preventing the fuel from leaking.

6.2   Micrometeorites and orbital debris impact detection

6.2.1   Methods not using fibre optic sensors

Different methods were proposed to detect the debris and their impact. Some use thin film sensitive to pressure variation [manganin and polyvinylidene fluoride] other AE, or from the variation of the electric current in an electric grid.

Tables 6.2 and 6.3 summarise the MMOD and debris impact detection methods not using optical fibres.

6.2.2   Detection with optical fibres

The detection of slow impacts using FOS goes back to the 1980s [5] with a system made of closely spaced optical fibres in an *X–Y* grid pattern embedded in a host material. The change of the intensity in fibres due to a small or medium impact or a crack can identify its location. The method was improved in more patents that followed the first version. Lymer *et al.* [6] proposed fibres with chemical etching to weaken them at certain locations so that they will break at predetermined strain levels. This system is only suitable to composites with known suitable for many applications because the failure strength of the fibres must be tailored for each case to ensure that the fibre will break at some acceptable damage threshold for the structure.

Fradenburgh *et al.* [7] proposed a system, based on untreated optical fibres, looped repeatedly in two separate *X*- and *Y*-direction grids before being embedded in or bonded onto a composite structure.

Table 6.2   Methods detecting hypervelocity particles and their impact

Detector	Comments
'Standard' laboratory gauges	Manganin thin-film gauge is commonly used in laboratories of shock waves and hypervelocity launchers to measure the impact pressure. It has low strain sensitivity but high hydrostatic pressure sensitivity. Thin-film manganin can have less than 1 ns response time. Manganin is an alloy (86% copper, 12% manganese, 2% nickel)
Piezoelectric	It is a strike detector used as thin sheets of PVDF plastic between two metal layers. A debris impact creates a voltage difference between the metal layers. This signal is measured by a central electronic unit that monitors all sheets covering the protected surface
AE	The AE probes emission is generated from active defects (permitting the observation of damage processes during the load history without disturbing the sample specimen). It differs considerably from ultrasonic probing which investigates rather the structure. The drawback of AE is its limitation to the quantitative estimation of the damage in the structure
Resistive grid	Resistive grid sensor is a passive dust particle-flux probing device. It is able to directly measure the particle size. The device contains about 1,000 resistive lines (75 μm wide/5 cm long) separated by gaps of 75 μm each. This specific design is aiming to detect debris larger than 50 μm. The wires diameter and distance can be changed as function of the addressed range of particle to monitor
Cameras	The method compares the differences between consecutive images or follows the photogram inspecting the impacted surface. The cameras can detect the damage side but need other sensor to evaluate the particle size and velocity. They are combined usually with other detectors
Accelerometers	Accelerometers are proposed for impact shock detection. They monitor the complex vibration environment generated by the impact, including in-plane and out-of-plane waves with different magnitude and frequency. Such disturbances propagate far away from the impact location, depending on the material, geometry, joints and other parameters
Thermography	It takes continuous sampling of the infrared radiation emitted by the local heat due to the impact and makes the difference with the previous sampling. Another method may require a suitable heat source to induce thermal waves inside the material, which are then sensed by an infrared camera. Differences in the structural cool-down response caused by thermal property changes may be finally associated to damages
Calorimetry	A main part of the debris kinetic energy is converted into heat after the impact, leading to a temperature increase. Using an energy absorber (e.g. metal with high thermal conductivity) with a temperature sensor permit to estimate the debris energy
Microwave emission	During the impact, a cloud is created with a local plasma including ions. The microwave emission by the plasma can be monitored using radio frequency antennas and pick-up foils [4]. This technology gives qualitative results and is combined with another detection method

Table 6.3   Combined methods detecting hypervelocity particles and their impact

PAE	The impact of a hypervelocity particle generates usually a relatively strong acoustic transient. This signal induces a strain when it goes through a material. The strain can be monitored easily with a piezo PVDF film. The frequency of the transient is in the 100 kHz range. There are different designs to consider depending on the structure, e.g. inflatable material, CFRP or metal. PAE sensor is used in different detection system by the NASA, Wing Leading Edge Impact Detection System on the Space Shuttle Orbiter as well as on ISS. DIDS concept combines PAE. It detects and locates impacts *via* a wireless sensors system. Current DIDS system concept is to detect leak locations on space vehicles and on ISS. The AE module is asleep until event signal threshold is crossed. The sensor module can record four signals at 1 MHz rate. Its batteries can last up to 5 years
Dual acoustic emission and resistive grid	DRAGONS NASA sensor: Acoustic sensors detecting impact time and location on the two layers. The difference in time between the two layers provides the speed of debris. The resistive grids damage size is related to the size of the impacting debris. A third plate can be added to measure the impact kinetic energy, which, when combined with data from the acoustic sensors and the resistive grid, provides a simple estimate of the density of the impacting particle
Cameras and laser ranging	NASA Space Shuttle uses an OBSS to scan the leading edges of the wings, the nose cap and other parts of the vehicle for impact damage soon after each lift-off and before landing. OBSS objective is to detect any potential critical damage to the thermal protection system of the vehicle caused by debris from launch and from orbiting. OBSS instrumentation packages include a laser dynamic range imager, an intensified television camera, a laser camera system and a digital camera
Calorimeter and optical	Combining calorimetry with an optical laser system has been explored through the development of the AIDA detector. AIDA enables the particle velocity measurement. Laser diodes produce a thin (~3 mm) light sheet through which particles impacting the spacecraft must travel. As they pass through the sheet, they scatter the laser light, which is picked up by a set of photodiode detectors. By using a set of two or more laser light sheets, the velocity of the particle can be determined by the difference in time and position of the light sheet crossings
Combined detectors	MIDS is a solid metal plate of 1×1 m. A thin film which is maintained under tension is placed at about 1 cm above the plate. The MIDS consists of an acoustic sensor to detect the signals generated by particle impacts with the bottom plate, and fibre optic displacement (FOMIS) sensors to measure the film' motion following the particles impacting or penetrating the film. SEDS: The basic configuration of SEDS includes a DOCS and the acoustic sensors that are attached to a solid plate behind the second optical curtain of the DOCS. The SEDS is configured to characterise the impact flux, velocity, size, mass and density of the detected secondary eject particles. DOCS is based on grain impact analyser and dust accumulator developed by ESA for the Rosetta mission

AIDA, advanced impact detector assembly; DIDS, Distributed Impact Detection System; DOCS, dual-layer optical curtain sensor; FOMIS, fibre optic micrometeoroid impact sensor; ESA, European Space Agency; MIDS, Micrometeoroid Impact Detection System; OBSS, Orbiter Boom Sensor System; PAE, piezoelectric acoustic emission; SEDS, Secondary Ejecta Detection System.

Lu *et al.* [8] employed a laser diode-driven system with two-mode elliptical core optical fibre as an impact and damage sensor. The need for closely spaced fibres is mitigated by the ability of the sensor to detect the effect of acoustic waves at distances of up to 1 ft. The coupling of a multiplicity of fibres into the laser diode is required to accommodate large structures and intensity loss makes this system not well practical.

Sirkis *et al.* [9] developed an optical fibre sensor coated with linear work hardening elastic-plastic materials and employed for a non-destructive optical damage detection sensor. The detection coating's intrinsic and extrinsic properties are critical on the sensor's continuous performance while being subjected to adverse environmental conditions. The damage detection sensor can also be used as an alternative to the conventional fibre fracture sensor for sensing impact damage. Other uses include maintaining a permanent record of the load/damage history of a loaded structure.

Furthermore, Sirkis proposed an improved system in 1994 – the fibre sensors for the debris impact [10]. The sensors locate the position of impacts and quantify the corresponding strain energy absorbed by the space structure. The techniques were complex, with sensors demodulation and neural networks for decision-making.

Pope *et al.* [11] proposed a fibre optic damage system that employs an optical time domain reflectometer to locate faults within long lengths of optical fibre; however, that system is directed to fault detection for electrical conductors.

Hirayama *et al.* and Morison *et al.* [12,13] proposed an improved FOS-based system to detect and quantify the area of damage caused by an impacting object and more particularly the area of damage caused by the impact of a hypervelocity object. The invention is further directed to a method for using such a sensor system. In his patent application, Tennyson presented a good overview of the fibre sensors previously proposed for detecting the impact of small debris.

Morison [13] proposed an FOS to measure the strain by the Fourier–Transform (FFT) interferometer equivalent to a Fabry–Pérot (FP) fibre sensor. He proposed a grid of 1,000 fibre sensors for the ISS, but there was no further follow-up. However, Tennyson relies on FP fibre sensors which have many limitations, including the fact to have only one sensor per line and the difficulty of the data analysis due to Fourier inversion of the FP interferometer.

Morison *et al.*, Nakamura *et al.* and Asanuma *et al.* [13,14,15] proposed fibre sensor based on the fibre Bragg gratings (FBGs) to monitor the debris impact on the space stations, as a general idea. No further development by these teams could be found.

By using embedded FBG sensors, a camera scan and microscopical analysis, Chambers *et al.* [16] assessed the damage occurring in composite material resulting from various levels of velocity/energy impacts. The FBG sensors were placed at 10 mm from the impact zone and were found to detect residual strains from a 0.33 J (1.3 m/s) impact. This threshold could not be detected either by C-scan or visual inspection. The measured residual strain was found to be more important as a function of the impact energy. The occurring damage onto the composite has also changed from a simple matrix cracking to severe and complex delamination. By increasing the impact energy further, high-velocity impacts, occurring at 225 m/s (i.e. 11 J), have resulted in the perforation and delamination of the test panel.

By using a network of embedded FBG sensors, it is expected that the identification of the site of both low- and high-velocity energy impacts becomes possible, and one can even predict the damage by analysing the response of the near neighbouring FBG sensors that are closely located to the impact site. Note that the velocities used by Chambers *et al.* [16] were below that of the space debris range. However they successfully demonstrated the possibility of detecting the residual stress after the impact event.

Kirikera *et al.* [17] proposed a fast data acquisition with multiplexed FBG sensor that has the capacity to measure dynamically ultrasonic wave frequency above 200 kHz using a spectral demodulator with a two-wave mixing that allows multiplexing of multiples FBG sensors at the same time and acts also as a filter with a high pass permitting the elimination of the low-frequency thermal drift together with vibrational noise. This 'two-wave mixing spectral demodulator' does not require any stabilisation as it is the case with other interferometric demodulators. The idea is creative and innovative; however, two-wave mixing is very challenging and sensitive to vibration on ground applications; for space it would be even more difficult.

On the other hand, FBG sensor technology was identified by NASA as the most promising future technology for structure monitoring [18]. The NASA reports propose FOS in two forms:

• A series of FBG distributed at known distance between them, the series can be a few hundreds, detecting the strain by their centre wavelength (CWL) shift at different position of a structure. The specific FBG positions are identified through optical frequency domain reflectometry (OFDR). The strain data can be correlated into displacement data providing the structure deformation. The FBG can measure vibration strain combined with the OFDR called fibre optic sensing system [18]. This method permits to convert directly the raw data into frequency/location information, hence acquiring the data in their final shape. The process is to divide the raw FBG sensor data into several segments over time [19,20].

• A second method based on intensity variation in the optical fibre grid with respect to the impact event [21]. The FBG sensors can also be interlaced into hybrid MMOD shielding fabrics. The glass fibres of the FBG sensor furnish a double aim in contributing to the breakup of MMOD projectiles. The grid networks can be made in a modular configuration to allocate coverage over any targeted area. Each module in the array can be connected to a central scanner instrument and be interrogated in a continuous or periodic mode. Table 6.4 summarises the MMOD-impact-detection methods using optical fibres.

6.3 Experimental study of micrometeorites and orbital space debris

Different launchers can be used in laboratories with pellets size from 0.5 mm up to 2.5 cm and hypervelocity up to 15 km/s. These launchers are employed for standard classification of materials response to hypervelocity pellets. The Inter-Agency Space Debris Coordination Committee from the UN has a list of laboratories with calibrated hypervelocity measurements. Due to the high launching cost, most of the

Table 6.4   Fibre sensor methods detecting hypervelocity particles and their impact

Passive fibre	Optical fibre grid with impact detected *via* optical signal intensity change, proposed by the NASA. The fibre can be coated with metallic or chemical product to break preferentially at well-known point
FBG	FBG can monitor the vibrations after impact. NASA developed a sensor system called FOMIS. This requires interrogators at 2–10 kHz acquisition frequency. Such interrogators are being qualified for space environment. The highest acquisition speed of FBG spectrum is provided by micron optics interrogator at 2 MHz. It can follow certain details of the impact effects
FBG and OFDR	Combining FBG measurements (strain, vibration) with OFDR (time, local FBG position) proposed by the NASA for many projects
Two-wave mixing	Using high frequency (>200 kHz) ultrasonic waves by two-wave mixing spectral demodulator to monitor high-frequency dynamic strains (>10 kHz)

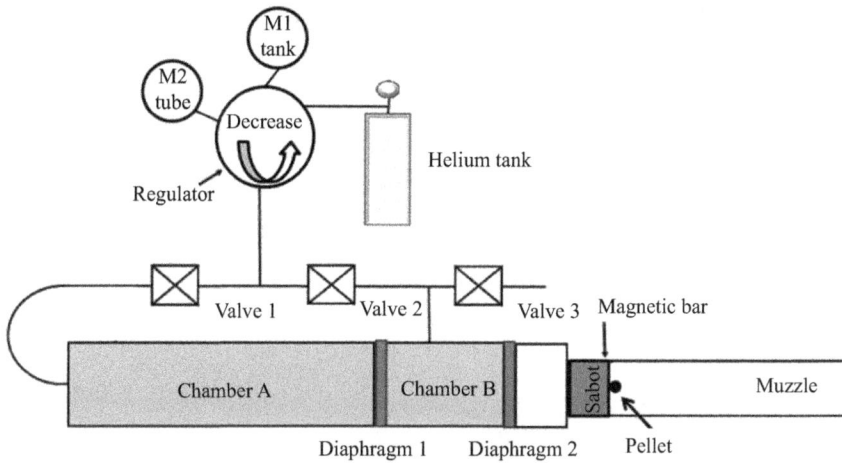

Figure 6.3   Schematic diagram of a two stages launcher

laboratories used simplified launchers as the double stages shown in Figure 6.3, which can get up to 4 km/s. The pressure handled by the diaphragm in chamber A/B can be twice higher than the diaphragm B/Tube. When the diaphragms break together, the pellet receives double the pressure compared to single-chamber launcher leading to higher hypervelocity. Figure 6.4 illustrates the details of the diaphragm and the pellet support (sabot) with the magnetic bar.

The pellet speed can be measured by different methods such as fast streak camera, laser diodes or fibre laser Doppler shift. One simple method to measure the speed is to insert a small magnetic bar (0.5 mm diameter, 1.5 mm length) in the lateral side of the sabot. Two coils are enrolled separately by a fixed distance (5–10 cm) close to the end of the launcher tube. The variation of the coils voltage in time when the magnetic bar passes is registered providing the hypervelocity measurement (Figure 6.5).

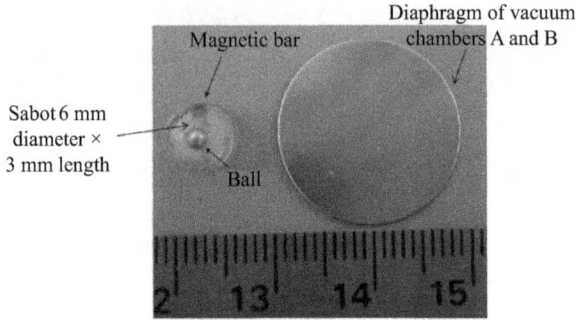

*Figure 6.4   Details of the diaphragm size and the pellet (ball) with its support
(sabot) and the magnet bar inserted in the lateral side of the sabot*

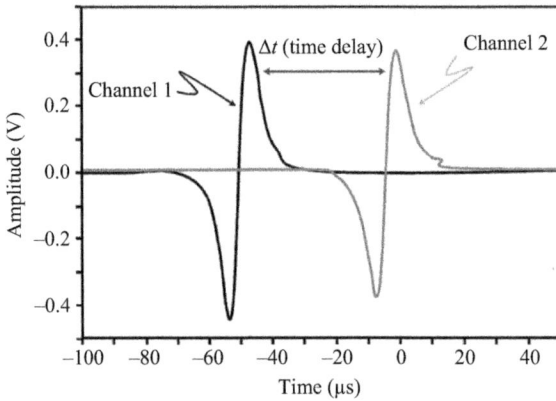

*Figure 6.5   Signal in the two coils separated by 5 cm, when the magnetic bar
signal passes between*

6.4   Fibre Bragg grating response to hypervelocity impacts

The FBG can sense the debris impact in different methods, in particular three are used:

1.   Continuous complete spectrum, with very fast interrogators, to extract the CWL shift due to the impact. The fastest interrogator commercially available is at 2 MHz with complete sweep between 1,520 and 1,610 nm (Micron Optics sm690). Such interrogators can be used in laboratory to calibrate the fibre response with standard strain gauges (SG); however, they are not suitable for harsh space environment. Figure 6.6 compares the response to a hypervelocity mechanical shock of a manganin gauge at 1 GHz acquisition frequency with an FBG sensor at 2 MHz. The acquisition at 2 MHz was found to not be fast enough to register the details of the shock.

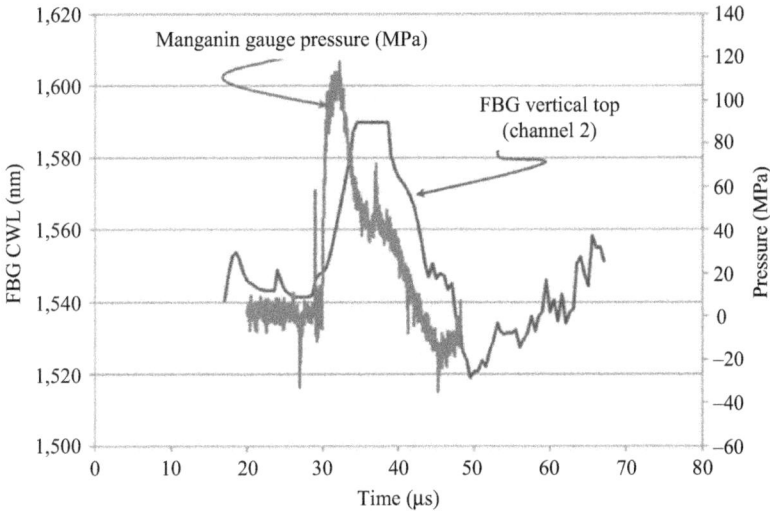

Figure 6.6   Comparison of the response of an FBG sensor and a manganin gauge to a hypervelocity mechanical shock

2. Very fast photodiodes (up to GHz, single-point acquisition), permit measuring the total intensity reflected by the FBG before, during and after the impact. Chirped FBG are being used to monitor the shock wave propagation, by measuring the wavefront position of the detonation and the changes of the velocity through the material interfaces [22]. The dynamic mapping of both the detonation front position and the velocity as a function of the time are obtained from the FBG intensity variation in time [23]. The method is based on the total light return using a broadband light source reflected from a linearly chirped FBG. The relationship between the position and the wavelength is usually linear, leading to a linear relationship between the part of the FBG destroyed and the intensity of the reflected light.

3. The complete FBG spectrum before and after the impact can be compared to provide the residual stress and its spatial distribution after the impact of different host materials. This is very useful, since it will permit to verify in space (e.g. ISS) using local slow interrogator module to monitor the debris impact on the external surface, through the change in the residual stress distribution.

6.5   Experimental results with different target materials

6.5.1   Main effects of space debris micrometeorites

Impacts occurring at hypervelocity lead to a shock wave in the material and providing a very high pressure (>100 GPa) and temperature superior to 9,726.85°C. Further-detailed supplementary information is given, for example, in the ESA space debris mitigation handbook of the reference [4]. However, the following points should be considered:

- The impact event only lasts for few microseconds.
- The impactor (i.e. material undergoing the impact) and the target material are then fragmented, sometimes molten or even vapourised, depending on the total energy of the impact (i.e. its velocity and weight) and the properties of the materials.
- Most of the impact energy ends up in the ejecta (i.e. ejected mass).
- The ejected mass can be much important than the mass of the impactor because it includes that of the target.
- Often, a small fraction of less than 1 wt% of the ejected material is ionised. This phenomenon is function of the impactor velocity.

On the other hand, collision damage (Table 6.1) depends on

- kinetic energy of the particle (speed)
- design of the spacecraft (bumpers, external exposure points)
- collision geometry (especially the angle of collision).

The impact ranges are of about:

- 1 cm (medium) at 10 km/s: this can fatally damage a spacecraft.
- 1 mm and less: This erodes thermal surfaces, damage optics and puncture fuel lines.

The near space orbital is typically highly polluted by significant footmarks of recent human space history. Table 6.5 illustrates the number of objects orbiting in space, together with their categories, sizes and the Probability of collision and expected effects. Figure 6.7 is an illustration of the spread of collision debris orbital planes [24]. All the spacecrafts that have left the earth are participating to this evolution of collision risks in space for active space vehicles. The population of space debris includes a very large variety of fragments, ranging from the smallest ones (i.e. below a millimetre) to a complete vehicles size (i.e. up to several metres for lost spacecraft) (Table 6.5).

The use of CFRP in space structures has largely spread, it can be noticed through the number of published works dedicated to study its reliability, health monitoring in space and its response to debris.

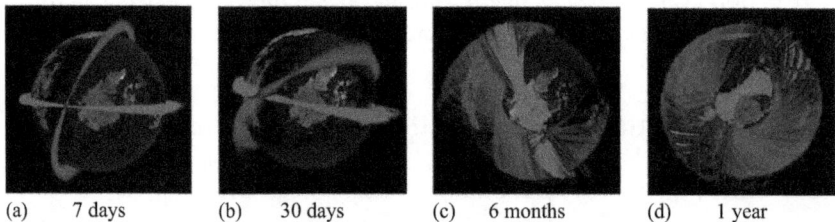

(a)   7 days   (b)   30 days   (c)   6 months   (d)   1 year

Figure 6.7   Spread of collision debris orbital planes (a) 7 days, (b) 30 days, (c) 6 months, and (d) 1 year after the collision. © 2009 NASA. Reproduced, with permission, from [24]

Table 6.5 Summary of a number of objects orbiting in space

Category (and origin)	Size	Numbers in orbit	Probability of collision and expected effects
Large debris: satellites, rocket bodies, fragmentation material	>10 cm	$1–5 \times 10^4$ (low) 17,800 in 2001	1/1,000 Collision results in total breakup and loss capability
Medium: fragmentation debris, explosion debris, leaking coolant	1 mm–10 cm	$1–5 \times 10^6$ (medium) 500,000 in 2001	1/100 Collision could cause significant damage and possible failure
Small: aluminium oxide particles, paint, chips, exhaust product, bolts, caps, meteoric dust	<1 mm	$>10^{12}$ (high) 3,000,000 in 2001	Almost 1/1 Collision could cause insignificant damage

Typical satellite service modules are square or octagonal boxes with a central cone/cylinder and shear panels (SP). The cone cylinder and SP are generally constructed from a sandwich panel with CFRP face sheets and an aluminium HC core. Similarly, the upper and lower platforms are also CFRP/aluminium HC SP. The lateral panels of the service module are, due to thermal reasons, generally sandwich panels with aluminium face sheets and aluminium HC cores. These panels are also generally wrapped with multilayer insulation blankets.

Lateral panels may be made with CFRP/aluminium HC SP. Other payloads include telescopes, which require a specific design quite often constructed predominantly of CFRP (for stability and pointing requirements). Truss-type structures are often used for supporting antennas, solar arrays, and so forth, typically made of CFRP.

6.5.2 Experimental work

Self-healing materials composite was fabricated at MBPC to mitigate the effect of the space debris. This composite was made with single-walled carbon nanotubes (CNTs) (SWCNT) materials, encapsulated monomer 5-ethylidene-2-norbornene (5E2N) and Grubbs' catalyst. In the following sections, we see a detailed experimental work achieved at our labs level.

SWCNT have been synthesised by using the plasma torch technology (detailed process can be found in [25,26]). Figure 6.8(a) shows representative transmission electron microscopy (TEM) micrographs of the purified SWCNT deposit.

6.5.3 Selection of the 5E2N monomer for the self-healing materials

All the chemicals (Grubbs' catalyst first generation, monomers and so on) were used as received. Due to the relatively high freezing temperature of dicyclopentadiene (DCPD), another monomer with larger functional temperature range was necessary. A number of candidates were evaluated, as shown in Table 6.6. Among these candidates, the monomer 5E2N is a liquid from $-80°C$ to $+148°C$, with faster ring-opening

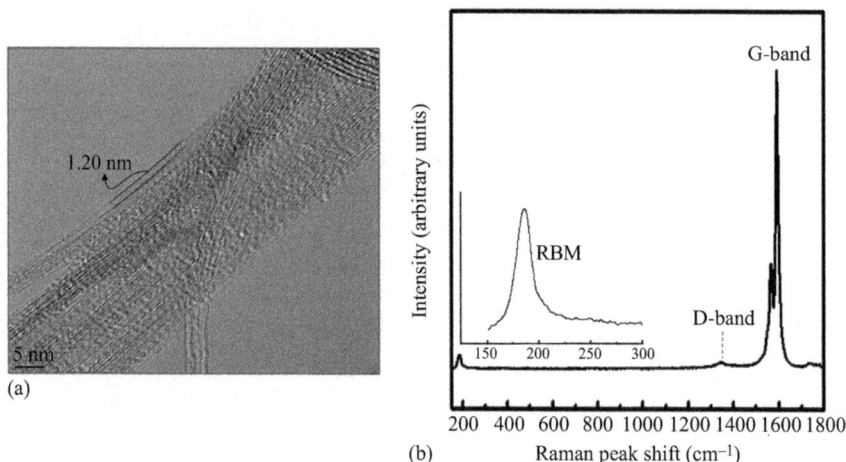

*Figure 6.8   Morphology of the grown SWCNT: (a) representative TEM images
of the SWCNT materials. The inset is a high-resolution TEM close up
showing nanotubes of 1.2 nm diameter. (b) Typical Raman spectrum
of the nanotube materials, where the various radical breathing mode
(RBM), D- and G-bands are clearly identified. The inset shows a
close-up of the RBM band located at 185 cm^{-1}, corresponding to
an individual SWCNT having a mean diameter of 1.2 nm in total
agreement with the TEM analysis in (a)*

metathesis polymerisation (ROMP) reaction that is DCPD [27,28]. Also, from the cost
and toxicity point of view, this monomer is more attractive than the others.

While 5E2N has a wider liquid temperature range than DCPD, it is important
to investigate its stability. The self-healing technique was tested in vacuum and
over a long time. To permit simple curing, the selected monomers for the self-
healing have to comply with the following requirements:

- Must be air stable
- Undergo a ROMP reaction

The study permitted a preliminary selection of the following compounds summarised in Table 6.7:

- DCPD
- Di(methylcyclopentadiene) (DMCP)
- 5E2N
- 5-Vinyl-2-norbornene (5V2N)
- 1,5-Cyclooctadiene (COD)

Hence, as mentioned above, the 5E2N was selected as the optimal monomer for
space environment with very wide range of temperature functionality [−80°C to +
148°C] and is less expensive and better relative to safety and environment (toxicity
and flammability).

Table 6.6 Comparison between different monomers

Compound	Toxicity	Melting temperature (°C)	Boiling temperature (°C)	$/L
DCPD	Flammable, harmful	32.5	170	~70
5E2N	Harmful	−80	148	~85
COD	Harmful	−69	150	~70
DMCP	Poisonous	−51	70–80	~60
5V2N	Harmful	−80	141	~160

Table 6.7 Shifts in wavelengths corresponding to different peaks observed from the same FBG after impact

Shifted peaks (nm)	Pressed Δλ (nm)	Strain (Δε)	Shifted peaks	Extended Δλ (nm)	Strain (Δε)
1,549.3	−2.1	−0.00173	1,551.4	0	0
1,549.8	−1.6	−0.00132	1,552.4	1	0.000823
1,550.1	−1.3	−0.00107	1,552.8	1.4	0.001152
1,550.3	−1.1	−0.00091	1,553.3	1.9	0.001564
1,550.6	−0.8	−0.00066	–	–	–
1,550.9	−0.5	−0.00041	–	–	–
1,551	−0.4	−0.00033	–	–	–
1,551.3	−0.1	−8.2E−05	–	–	–

6.5.4   The 5E2N monomer encapsulation

For the fabrication stage, the first issue is to achieve a controlled microencapsulation synthesised with desired characteristics (e.g. size, shell thickness, healing content and so on) and desired properties. The second issue is to disperse the microcapsules as well as the catalysts, appropriately into the matrix materials (Epon^{TM} 828 with EPICURE 3046), and finally making the sample. The samples are then tested by creating certain damages into it, and healing performance is observed.

The encapsulation of the 5E2N in poly(melamine-*urea*-formaldehyde) (PMUF) microcapsules was achieved following the protocol here below (Figure 6.9). The drying of the microcapsules was improved by rinsing them successively with deionised water and acetone.

Microcapsules are synthesised according to the flow chart in Figure 6.9. The process is inspired from the one described in [29]. Several batches of microcapsules were produced. The size of the microcapsules was mainly controlled by varying the stirring speed during the synthesis process. All samples were dried in air for at least 24 h after their final washing and filtering. Separating the microcapsules from the dross is found to be the main challenge of the process. Figure 6.10(a) shows typical optical micrographs of microcapsules produced with stirring speed of 1,200 rpm. However, a few microcapsule shells are found to break (Figure 6.10(b)), which is possibly due to the high current/voltage used during the scanning electron microscopy (SEM) observation.

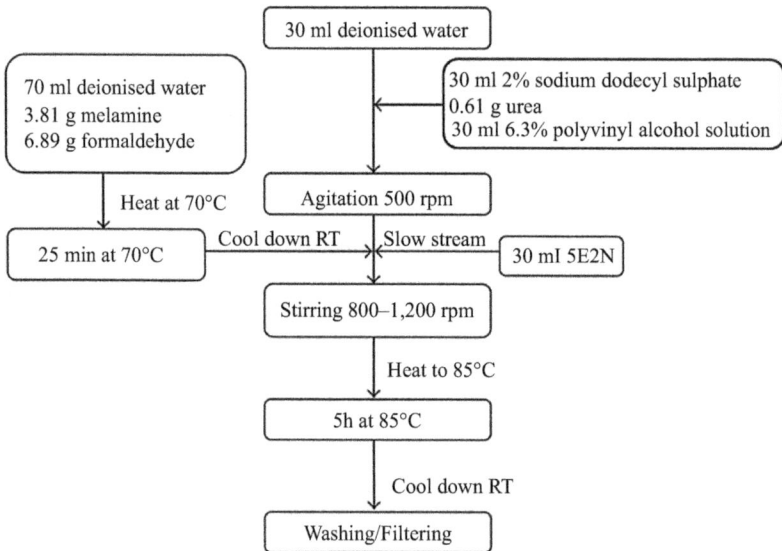

Figure 6.9   Flow chart for preparing 5E2N/PMUF microcapsules (RT, room temperature)

(a)   (b)

Figure 6.10   5E2N monomer encapsulation (a) typical optical photo, followed by (b) SEM image of the small microcapsule showing the core-shell structure of the encapsulation

6.5.5   Chemical constituents

The self-healing demonstrator consisted of woven CFRP samples, containing four main constituents:

- Host matrix: An epoxy prepolymer (Epon 828) and a curing agent (EPICURE 3046); this epoxy is used in space for internal structures.
- Microcapsules: The monomer healing agents (5E2N) prepared as small microcapsules (diameter less than 15 μm); the monomer is homogeneously spread within the epoxy and forms about 10% of the structure weight.
- Different concentrations of SWCNT materials.
- Catalyst: Grubbs' catalyst first generation (ruthenium metal catalyst).

6.5.6   Monomer diffusion under vacuum

The diffusion of the monomer from the pure microcapsules in vacuum was measured by placing a weighted amount of capsules in a Schlenk tube under vacuum (10^{-2} Mbar) for an extended period of time. We see from Figure 6.11 that the histogram presenting a plateau in the weight loss is reached after 5 days (34 wt%, i.e. about 50% of the initial monomer content of the microcapsule is lost). The monomer loss seems to proceed by steps.

6.5.7   Fabrication of woven carbon-fibre-reinforced polymer samples embedded microcapsules containing 5E2N and dicyclopentadiene monomers and Grubb's catalyst

The CFRP samples containing self-healing demonstrator consists of epoxy used in space for internal structures (Epon 828 resin, with the EPICURE 3046 curing agent) and two different healing agents (namely, 5E2N and DCPD) prepared as small microcapsules (diameter less than 15 μm) kept within thin shells of PMUF; according

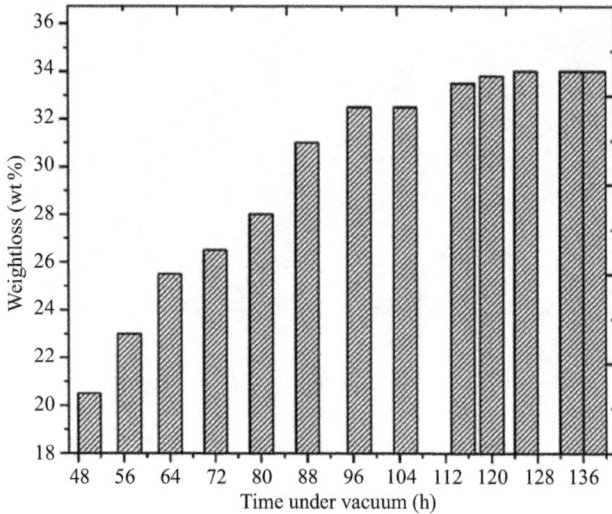

Figure 6.11   Microcapsule weight loss percentage under vacuum as a function of time

to the flow chart of Figure 6.9. The monomer is homogeneously spread within the epoxy and forms about 10% of the total weight. The Grubbs' catalyst was then distributed within the epoxy structure (1%–2% of the total weight). Different series of samples specimens were prepared, with and without CNT. After the hypervelocity impact tests, the crack formed on the CFRP samples reaches a microcapsule and causes its wall rupture, which releases the healing agent monomer (5E2N or DCPD or the combination of the two monomers as will be detailed below) in the crack. Once the monomer and catalyst enter in contact, the self-healing reaction is triggered (i.e. polymerisation between the healing agent, monomer, and matrix-embedded catalyst particles) [29]. The Grubbs' catalyst will sustain the ROMP chemical reaction known up to the time the crack and the open microcapsule are full.

The use of the 5E2N monomer can be considered as an innovative healing method, since none of the chemical products proposed previously would be functional, in space environment conditions, because the DCPD monomer melting temperature is 33°C (i.e. it is in the solid phase at RT), and the chemical activities are somehow relatively slow, with the risk of evaporation during the reaction. However, the drawback to use 5E2N monomer is that the formed polymer is linear and has consequently low mechanical properties as compared to the polymer issued through DCPD monomer.

A route towards the combination of these two monomers (Figure 6.12) was proposed, thereby obtaining simultaneously a fast autonomic self-repair composite with excellent mechanical properties. A second step aimed at combining the known higher mechanical properties of CNT materials with these monomers to evaluate the self-healing capability.

This section presents examples of FBG monitoring the impact on different composite materials with hypervelocity pellets. The two types of FBG are used: (i) the common sensors having a single CWL and about 0.5 nm width, and a size

```
┌─────────────────────────────────┐
│  Dissolve the Grubbs' catalyst (5 mg) into │
│  10 ml of acetone               │
└─────────────────────────────────┘
                                        ⇓
┌─────────────────────────────────┐
│  Blend with epoxy (Epon™ 828).  │
│  Manual mixing                  │
└─────────────────────────────────┘
                                        ⇓
┌─────────────────────────────────┐
│  Blend with microcapsules:      │
│  10 wt% microcapsule and X wt% SWCNT. │
│  Vacuum mixing                  │
└─────────────────────────────────┘
                                        ⇓
┌─────────────────────────────────┐
│  Mix with curing agent (EPICURE 3046). │
│  Note: Epon™ 828/EPICURE are 100/40 │
│  parts. Vacuum mixing           │
└─────────────────────────────────┘
                                        ⇓
┌─────────────────────────────────┐
│  Incorporating the mixture into the CFRP │
│  layers (between each of the two layers). │
│  Cure at 40°C for 10 h under 40 psi into the │
│  autoclave machine              │
└─────────────────────────────────┘
                                        ⇓
┌─────────────────────────────────┐
│  Post-cure at 40°C/24 h under air │
└─────────────────────────────────┘
```

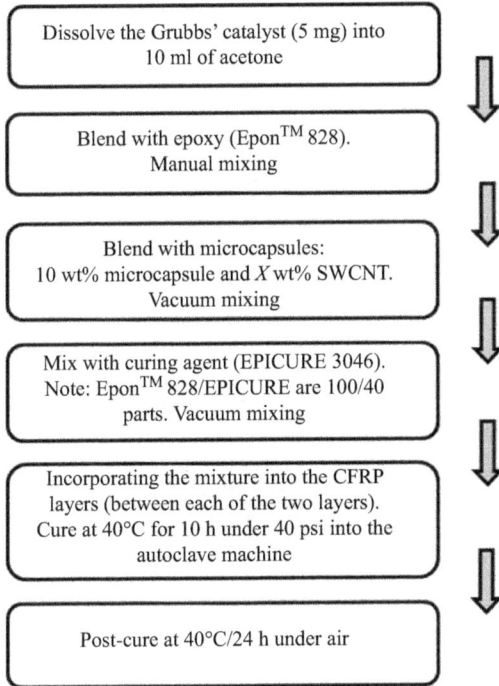

Figure 6.12   Fabrication procedure of the manufacturing of CFRP with self-healing and FBG sensors

are between 2 mm and 1 cm and (ii) the chirped sensor covering a wide wavelength range of about 30–40 nm with a linear relationship between the wavelength and its spatial position on the sensor. The chirped FBG can cover a spatial size of 2–5 cm, although extreme cases such as 1 and 10 cm long can be manufactured.

The target materials are CFRP or Kevlar laminates with Epon 828 between the layers. Some samples included within the resin microcapsules containing self-healing material. The CFRP laminates are being used in various airplanes and satellites, the Kevlar is a strong material proposed as one of the best resistant against debris.

6.5.8   Fibre Bragg grating within multilayer Kevlar/Epoxy

A single-wavelength fibre sensor is embedded in the middle of four layers of Kevlar with Epon 828 resin in between, a manganin SG attached at the back side (Figure 6.13). The resin contained 1% CNT for reinforcement (1% total resin mass) and self-healing microcapsules (10% of total mass). The SG signal and the total intensity signal of the FBG are monitored at 200 MHz acquisition rate. It took 250 ns for the pellet to go across the half of the sample thickness (Figure 6.14). The pellet was a 2 mm stainless steel with a speed of 1,050 m/s.

Figure 6.15 illustrates the spectra of the single-wavelength FBG before and after the impact, showing different points of different response on the 1 cm length of the sensor. Moreover there is some self-repairing with time showing the strain

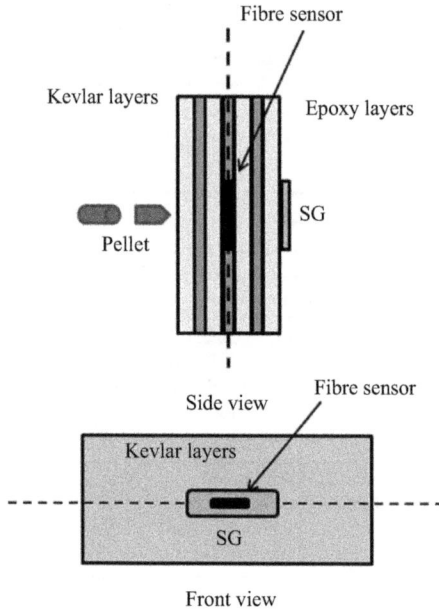

Figure 6.13   Schematic illustrating four layers Kevlar and epoxy, the fibre sensor is in the middle and the SG on the back side

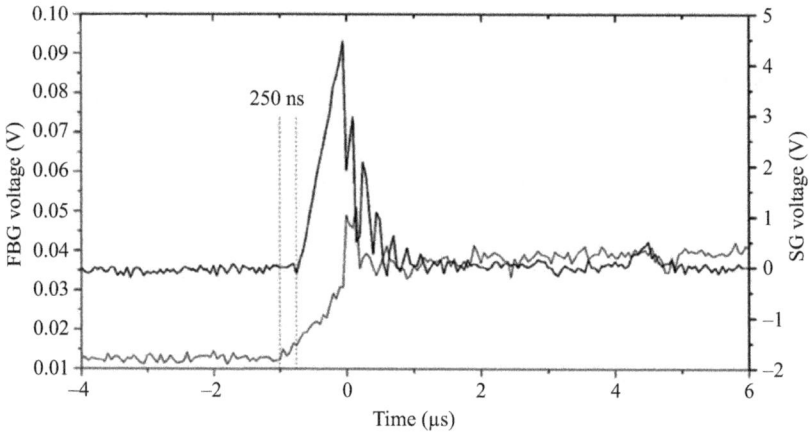

Figure 6.14   Fast acquisition of single-wavelength FBG sensor and SG

peaks are reduced with time, with a small peak emerging at the original CWL. Table 6.7 summarises the strains values related to the different observed peaks.

Figure 6.16 presents the fast acquisition of a chirped grating and SG for a sample of four layers Kevlar with resin but without self-healing microcapsules or CNT. It took about 430 ns between the response of the FBG and that of SG

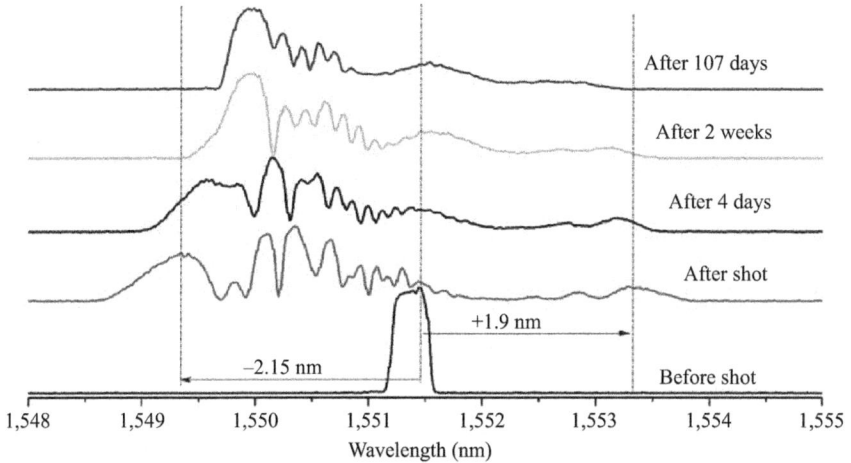

Figure 6.15   Evolution of the FBG–CWL spectra (negative-wavelength shift for stress; positive for extension)

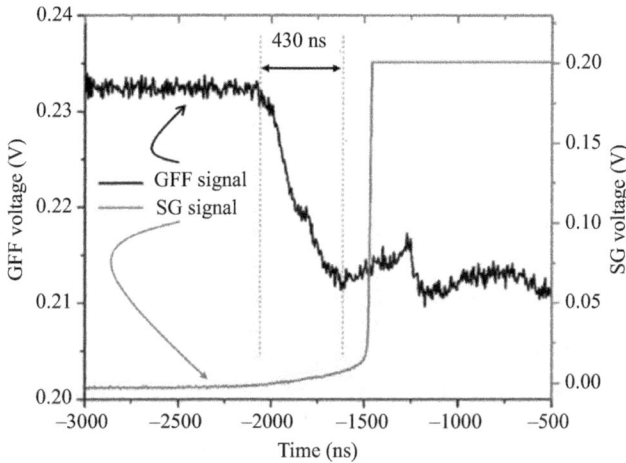

Figure 6.16   Fast response in time of total intensity of a chirped FBG gain flattening filter (GFF) and a SG

response. Figure 6.16 illustrates the chirped FBG spectra before the impact, then after impact, the reflection from the two sides of the fibres. Figure 6.17 presents the detailed spectra before and after the impact, in water shell view. The FBG total length is 25 mm. We can see the part broken and destroyed by the impact at about 1.2 mm. The two edges with about 6 mm are practically not affected by the impact. There is a well-localised residual stress or deformation in the FBG with 1–1.5 mm long, two at the left side and a third at the right side. There is a small partial self-healing as the spectrum after 86 days is coming back towards the original one.

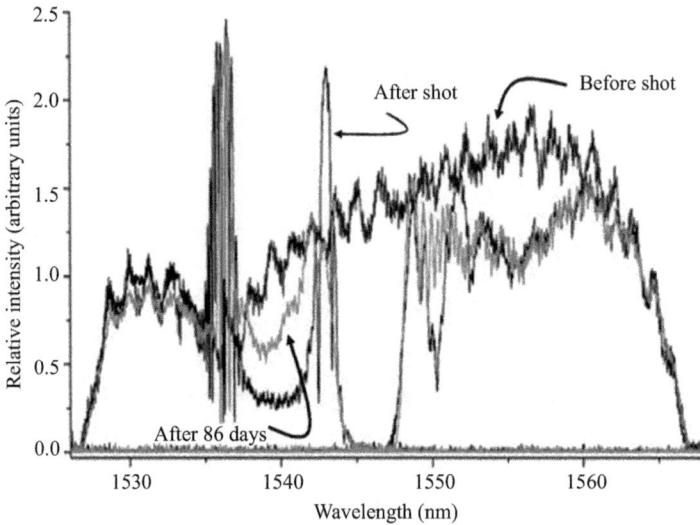

Figure 6.17   Chirped-FBG reflection before and after the impact

6.5.9   Fibre Bragg grating within multilayer carbon-fibre-reinforced polymer/epoxy

The CFRP/epoxy system consisted of eight layers CFRP with resin, some of them included self-healing microcapsules or CNT, and was embedding up to four fibre sensors (Figure 6.18). The Large Launcher at McGill University was used (Figure 6.19). The interrogator was a micron optics sm690 at 2 MHz acquisition rate.

We focused in this section mainly on the projectile velocity measurement. Three FBG sensors were placed on the implosion tube (Chapters 1, 2 and 3, respectively) and one FBG (channel 0, CH 0) on the CFRP. All sensors were interrogated with the high-speed 2 MHz system. The distance from the FBG3 to FBG0 is 300 cm. Knowing the first response time of these FBG (Figure 6.20), we can calculate the pellet hypervelocity. It was similar to that measured by the fast camera.

The objective was in addition to the healing efficiency:

- Measure the pellet velocity, from the fibre sensor signal, and compare to that measured by the streak camera (standard instrument). The camera measured a velocity of 7.9 km/s.
- Measure the strain evolution on the launching tube, to be compared with shock wave measurement.

The pellet velocity was calculated from the time difference of the FBG based on the pattern of the obtained signal (Figures 6.21 and 6.22), the response of the measured FBG–CWL signals as a function of the time. CWL were obtained for the impacted CFRP structures. For example, velocities were calculated based on the time delay of the change on the wavelengths, the one seen after the first observed jump of the FBG

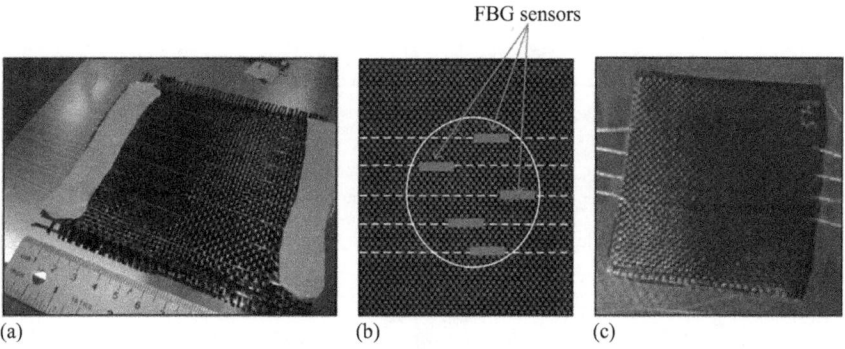

FBG sensors

(a)   (b)   (c)

Figure 6.18   (a) Integration of 4–8 FBG sensors embedded between second and third CFRP layer and concentrated inside a circle surface of 5 cm diameter corresponding to the exposed area to debris during the hypervelocity impact experiment, (b) position of the FBG sensors onto the optical fibres, and (c) final prototype of the sample

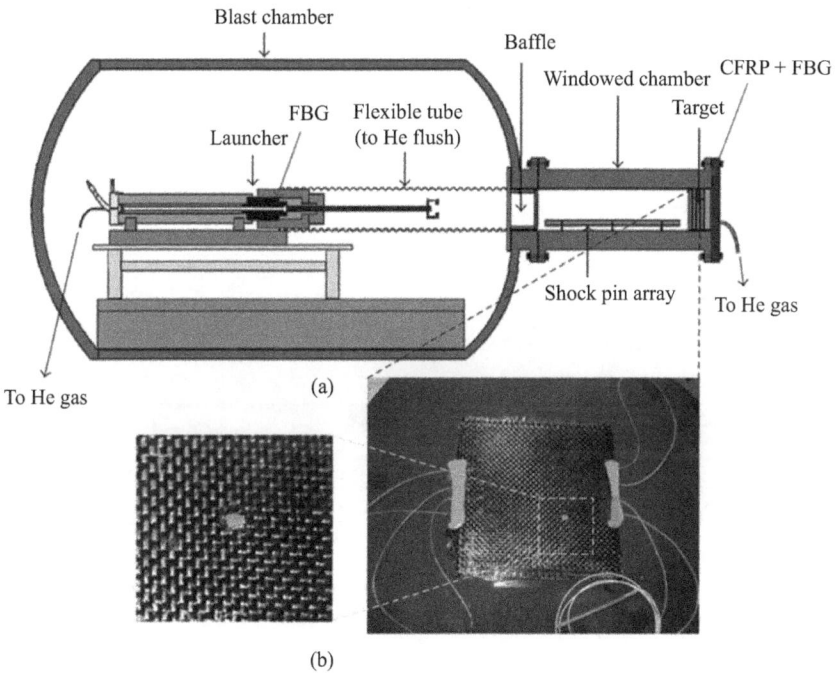

Blast chamber

Baffle

Windowed chamber   CFRP + FBG

FBG   Flexible tube   Target
Launcher   (to He flush)

To He gas   (a)

Shock pin array   To He gas

(b)

Figure 6.19   (a) Schematic of the launcher and the FBG. Some FBG were installed on the explosion tube (launcher, 7–8 km/s), and some at the end of the tube within the CFRP sample and (b) typical example of the CFRP samples containing FBG sensors after shooting under the hypervelocity impact

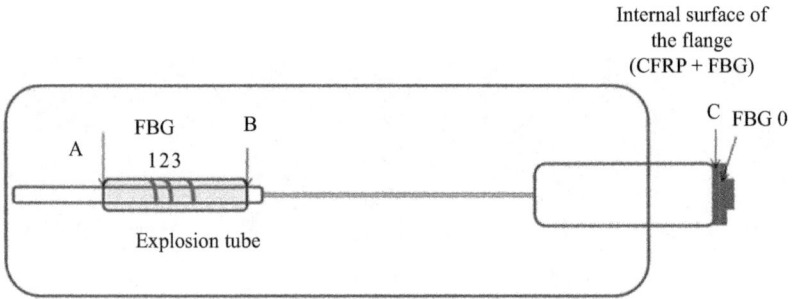

Figure 6.20   CFRP sample containing three FBG on the launching tube, one FBG within the CFRP

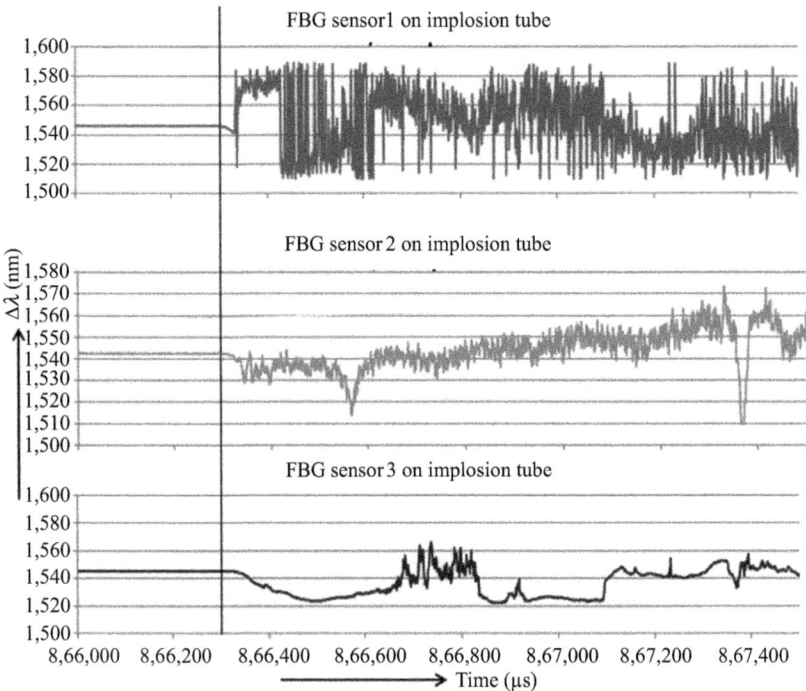

Figure 6.21   Typical recorded FBG signals obtained on the launcher implosion tube. The fibres broke after about 1 ms

signals. The goal is to have a good statistical data set for the velocity estimation. In sum, the embedded FBG sensors have shown a good capability to estimate both the impact event and the pellet hypervelocity, with accuracy within 1% in some cases.

One additional advantage of the fast acquisition is the possibility of the fast FFT to detect special resonance frequencies. Figure 6.23 shows the resonance frequencies between 90 and 110 kHz of four FBG sensors embedded in the middle of eight layers

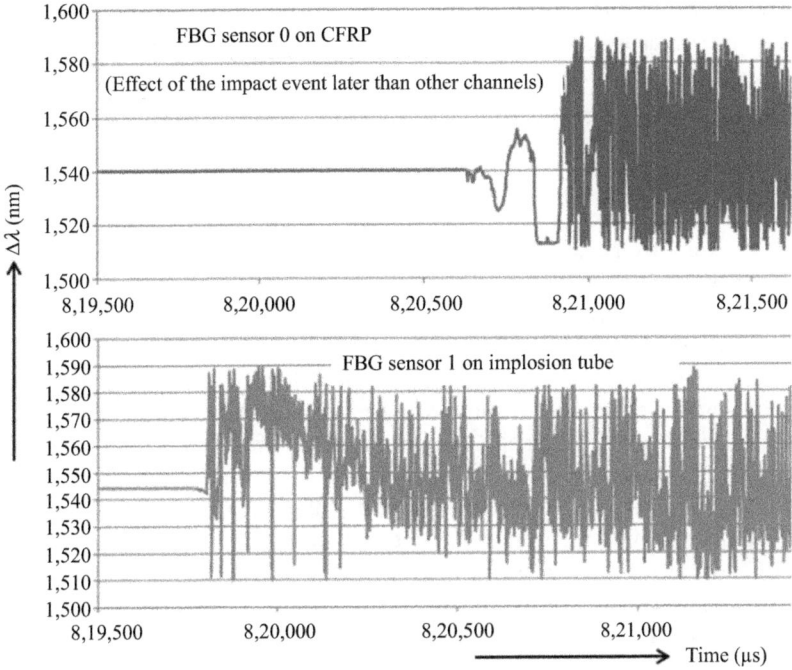

Figure 6.22 Representative FBG responses on the implosion tube and CFRP sample which occur about a half ms after the implosion event

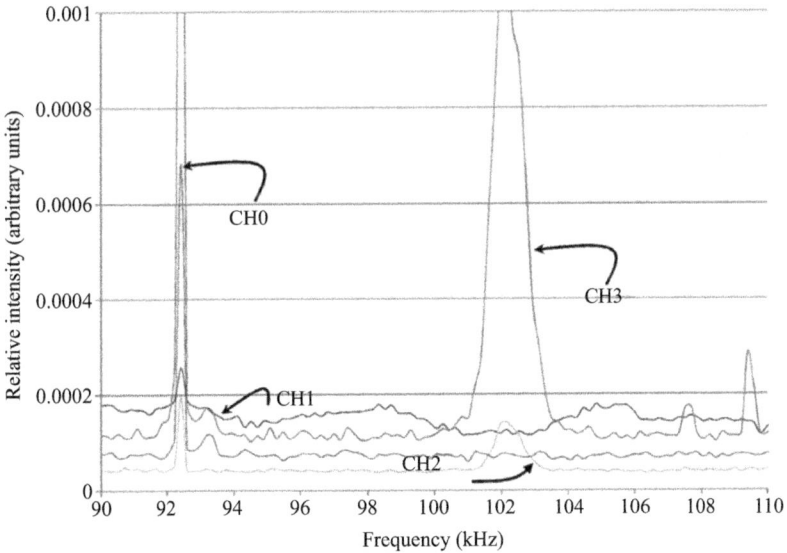

Figure 6.23 Frequency distribution (FFT) of the four FBG sensors embedded within CFRP layers

CFRP at 1.5 cm distance one from the next neighbour. We can see there are some common frequencies while others are different and related to a specific FBG.

6.6   Conclusions

This chapter summarised the classical and innovative methods of protecting space-craft or satellite components against small space debris. It also presented the detection and sensing methods proposed to monitor the impact and its characteristic.

The FBG sensors are very promising to monitor the micrometeorites and small space debris starting from their impact and fast effect with the possibility of comparing the effect at long term after the impact which common SG cannot perform. In addition, the size of the fibres permits to embed them or at least to attach them to a specific surface to monitor without major physical interference.

The FBG sensors potentially permit the detection of the location of micrometeoroids or orbital debris impact, the direction from which the impact occurred, the size of the impacting object and the extent of the damage caused by the impact.

FBG and standard SG were embedded in representative protective multilayers, the FBG demonstrated their fast response to the impact. Their signal can be easily correlated with that given by other sensors, to measure, for example, the shock propagation speed within the target. Although, there is no adequate commercial space-qualified FBG sensor system currently available, the technology is very promising.

References

[1]   J. Arnold, A. Davis, B. Corsaro, *et al.* in *Handbook for Designing MMOD Protection*, Ed., E.L. Christiansen, JSC-64399, NASA Johnson Space Center, Houston, TX, USA, 2009. *https://ntrs.nasa.gov/archive/nasa/casi.ntrs. nasa.gov/20090010053.pdf.*

[2]   B. Aïssa, K. Tagziria, E. Haddad, *et al.*, *ISRN Nanomaterials*, 2012, **2012**, 1–16.

[3]   W.L. Richards, E.I. Madaras, W.H. Prosser and G. Studor in *NASA Applications of Structural Health Monitoring Technology*, NASA Document ID: 20140010525, 2013. *https://ntrs.nasa.gov/search.jsp?R=20140010525.*

[4]   *IADC Protection Manual*, Inter-Agency Space Debris Coordination Committee: IADC Space Debris Mitigation Guidelines. Issued by Steering Group and Working Group 4. September 2007, 1–10. *https://www.iadc-online.org/ index.cgi?item=docs_pub*

[5]   S. Malek and B. Hofer, inventors; Airbus Defence and Space GmbH, assignee; US4603252A, 1986.

[6]   J.D. Lymer, N.D. Glossop, W. Dayle Hogg, R.M. Measures and R.C. Tennyson, inventors; Innovations Foundation of University of Toronto, assignee; US4936649A, 1990.

[7]   E.A. Fradenburgh and R. Zincone, inventors; United Technologies Corp., assignee; US5015842A, 1991.

[8] Z.J. Lu and F.A. Blaha, inventors; CMC Electronics Inc., assignee; US5072110A, 1991.

[9] J.S. Sirkis, inventor; University of Maryland, assignee; US5245180A, 1993.

[10] J.S. Sirkis, J.K. Shaw, T.A. Berkoff, A.D. Kersey, E.J. Friebele and R.T. Jones, *Proceedings of Smart Structures and Materials 1994: Smart Sensing, Processing, and Instrumentation*, Orlando, FL, USA, 13–14 February 1994, doi:10.1117/12.173943.

[11] R.E. Pope, Jr., K.S. Watkins and S.J. Morris, inventors; ELECTRO-TK Inc., assignee; US6559437B1, 2003.

[12] H. Hirayama, T. Hanada and T. Yasaka, *Advances in Space Research*, 2004, **34**, 5, 951.

[13] W.D. Morison, R.C. Tennyson and T. Cherpillod, inventors; Fibre Optic Systems Technology, assignee; US7189959B1, 2007.

[14] M. Nakamura, Y. Kitazawa, H. Matsumoto, *et al.*, *Advances in Space Research*, 2015, **56**, 3, 436.

[15] H. Asanuma, K. Ichikawa and T. Kishi, *Journal of Intelligent Material System and Structures*, 1996, **7**, 301.

[16] A.R. Chambers, M.C. Mowlem and L. Dokos, *Composites Science and Technology*, 2007, **67**, 6, 1235.

[17] G. Kirikera, O. Balogun and S. Krishnaswamy, *Proceedings of the 4th European Workshop on Structural Health Monitoring*, Shanghai, China, 25–28 October 2008, 840.

[18] W.H. Prosser in *Development of Structural Health Management Technology for Aerospace Vehicles*, NASA Document ID: 20040003713, 2003. *https://ntrs.nasa.gov/search.jsp?R=20040003713*.

[19] A.R. Parker, inventor; NASA, assignee; US8700358B1, 2014.

[20] P.J. Harmony and A.R. Parker, inventors; NASA, assignee; US9444548B1, 2016.

[21] S.L. Rickman, W.L. Richards, E.L. Christiansen, A. Piazzac, F. Pena and A.R. Parker, *Procedia Engineering*, 2017, **188**, 233.

[22] S. Gilbertson, S.I. Jackson, S.W. Vincent and G. Rodriguez, *Applied Optics*, 2015, **54**, 3849.

[23] G. Rodriguez, R.L. Sandberg, Q. McCulloch, S.I. Jackson and S.W. Vincent, *Review of Scientific Instruments*, 2013, **84**, 015003.

[24] L.J. Nicholas in *NASA Green Engineering Masters Forum*, NASA, San Francisco, CA, FL, USA, 2009.

[25] O. Smiljanic, B. Stansfield, J. Dodelet, A. Serventi and S. Desilets, *Chemical Physics Letters*, 2002, **356**, 3–4, 189.

[26] O. Smiljanic, F. Larouche, X. Sun, J. Dodelet and B. Stansfield, *Journal of Nanoscience and Nanotechnology*, 2004, **4**, 8, 1005.

[27] B. Aïssa, E. Haddad, W. Jamroz, *et al.*, *Smart Materials and Structures*, 2012, **21**, 10.

[28] B. Aïssa, R. Nechache, E. Haddad, W. Jamroz, P. Merle and F. Rosei, *Applied Surface Science*, 2012, **258**, 24, 9800.

[29] X. Liu, X. Sheng, J. Lee and M. Kessler, *Macromolecular Materials and Engineering*, 2009, **294**, 6–7, 389.

Chapter 7

Fibre sensors for space applications

Spacecraft systems require extensive *in situ* systematic control of their performance, both during ground validation and operation in the space environment. This is presently achieved through various electronic sensors at the cost of consequent mass and substantial performance penalty due to the shielding requisite. In order to reach the required high technological readiness of a system to be able to pass space qualification tests, complex research and development (R&D) processes has to be carried out. Due to the high costs of bringing and operating systems in space, the risk for a potential failure of the system has to be reduced to a minimum.

Several fundamental physics experiments and future trends for existing space technologies based on photonics are already under development or have been proposed. This chapter gives an overview of state-of-the-art measurement systems based on frequency standards and optical fibre sensing (OFS) with regard to satellite interrogation system.

7.1 Optical fibre sensing

In spite of the spectacular R&D advancements in the field of optical fibres, the fibre-based sensing is still not a standard technology for space applications. Although the advantages of OFS in spacecraft have been theoretically and practically studied in several space-related projects [1], however, until now, sensing is dominated by electrical means. In particular, thermal mapping with high number of temperature sensors is one application wherefore fibre-based sensing would enhance the measurement in terms of accuracy, reduced complexity and weight budget. The higher maturity electrical technologies compared to optical ones required for OFS in space has to be overcome by further development.

Several studies have already been carried out having in view the utilisation of OFS in space [1,2]. Up to now, only one fibre Bragg grating (FBG) interrogator has been developed for space flight demonstration on board the European Space Agency's (ESA) PROBA-2 (Project for On-Board Autonomy) satellite [3,4] (this will be detailed in the subsequent section). This interrogator provides six external sensor channels. Four of the channels support fibres that accommodate up to four FBG sensors each, the fifth and sixth channels read out one high temperature and one combined pressure/temperature (*P/T*) sensor, respectively. Since this system is used for feasibility demonstration on board a low Earth orbit (LEO) satellite where environmental loads and mission lifetime are low, it is built on commercial components.

7.2   Optical-fibre-sensing development for space

The main objective of the R&D focussed on OFS to identify the potential benefits of fibre optic instrumentation for launchers and to demonstrate the feasibility and functional performance with a representative interrogation demonstrator. Within this section, sensing based on fibre optic sensors (FOSs) is outlined.

7.2.1   Fibre Bragg grating sensors

As detailed in the precedent chapters, the most common types of FOS are FBG named after the British physicist William Lawrence Bragg. FBGs are in principle dielectric filters that are based on multiple layers of alternating high and low refractive indexes (RI). At every optical transition between two consecutive layers, light is reflected partially. When the optical path length differences of the reflections are integer multiples of the wavelength, constructive interference occurs. Thereby, FBGs are wavelength-sensitive filters that, by variation of layer thicknesses, can be designed to reflect one specific wavelength and transmit all others.

In optical fibres, the Bragg grating can be generated by inscription with an ultraviolet (UV) laser [5–7] (Chapter 5). In summary, the laser beam is split into two arms and is brought to interference at the position of the fibre core. Thereby, the cores meet at places where high-intense constructive interference occurs. This changes the RI of the core and results in a periodic grating with a period 'Λ' (the principles were detailed in Chapters 1–3).

FOS provides a promising alternative to electric-sensing technologies. FOS instrumentation has the potential to improve conventional instrumentation in several ways. The design and development activities prove the potential of FOS instrumentation in terms of the following:

- Capability of sensor distribution. Sensor arrays, i.e. several sensors with different grating periods inscribed along one single fibre can be used. The sensors reflect light at different wavelengths. OFS thereby can be regarded as a sensor bus comparable to electric sensor buses like one wire or I2C (inter-integrated circuit). Figure 7.1 illustrates the possibility of sensor distribution.

- Insensitivity to electromagnetic (EM) interferences (EMI). FBG sensors are readout optically; the measurement accuracy is not influenced by any EM fields near the sensor location. Furthermore, FOS does not induce any interference that could harm surrounding equipment wherefore EM shielding can be omitted.

- Reduced mass and volume of the sensor network. One single optical fibre is required that guides the light in both directions, from interrogator to the sensors and back. Optical sensor fibres can be embedded in composite materials which reduces the effort for assembly and integration because sensor mounting is not required.

The wavelength itself which is reflected by an FBG does not allow any conclusion about the current measurement parameter (e.g. temperature or strain) at the location

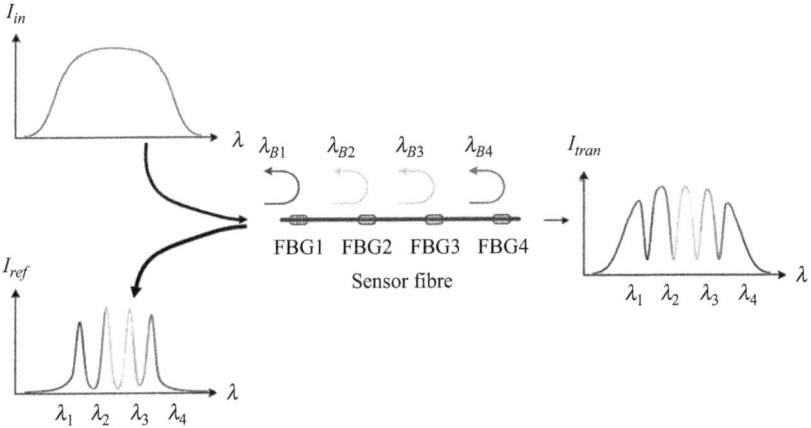

Figure 7.1   FBG with different grating periods that are inscribed in an optical fibre serially implement a sensor array. I_{in}, I_{ref} and I_{tran} are the intensities of the incident, reflected and transmitted optical signal having λ wavelength

of the sensor. Everything that influences the Bragg condition (Equation (7.1)) results in a change of the wavelength that is reflected. This wavelength shift is evaluated by the interrogator and allows to conclude about the relative change in magnitude of the actual measurand.

FBGs are sensitive to temperature and strain variations because both measurand changes influence the geometry of the FBG structure and the RI of the core material. According to [8], the spectral sensitivities of an FBG sensor at a design wavelength λ_B to applied strain ε and temperature T amount to (7.1) and (7.2), respectively:

$$\frac{\delta\lambda_B}{\delta\varepsilon} = 0.78 \times 10^{-6}\lambda_B/\mu\varepsilon \tag{7.1}$$

$$\frac{\delta\lambda_B}{\delta\varepsilon} = 0.78 \times 10^{-6}\lambda_B/K \tag{7.2}$$

In the case of a design wavelength of $\lambda_B = 1{,}550$ nm, the overall wavelength shift λ_B of an FBG due to applied strain $\Delta\varepsilon$ and temperature ΔT results in

$$\Delta\lambda_B = 1.21 \text{ pm}/\mu\varepsilon \cdot \Delta\varepsilon + 10.34 \text{ pm}/K \cdot \Delta T \tag{7.3}$$

One issue that arises from the sensitivity of an FBG sensor given by (7.3) is that it cannot be distinguished if the wavelength shift arises from a temperature or a strain variation. Similar to electric strain sensors, the sensor transducer needs to be designed such that unambiguity of measurement results is given. Transducers for temperature sensors, for example, have to decouple the sensor from structural strain.

7.2.2   Fibre Bragg grating interrogation systems

Three well-known configurations of fibre optic interrogation systems are commonly used, namely spectrometer-based systems that utilise a broadband illumination and detect the actual wavelength of the sensor by wavelength-division multiplexing (WDM). A second principle uses edge filters to 'cut' the sensor response and determine the measurement value by the ratio of the two resulting intensity signals. Finally, systems based on tunable lasers sweep through the wavelength spectrum and evaluate the magnitudes of the reflected (or transmitted) intensities.

7.2.2.1   Spectrometer-based systems

This design uses a broadband light source such as superluminescent diodes in order to illuminate FBG sensors within the sensor channels. The different sensors reflect light at different wavelengths. The reflected signals are spectrally encoded by a spectrometer consisting of, e.g., a diffraction grating and a charged coupled device (CCD) sensor. Electronics evaluate the CCD image whereof the mean wavelengths of all sensors are determined by centroid algorithms (Figure 7.2).

This kind of interrogator concept has the highest maturity; all components are commercially available as standard industrial components. By enhanced design of the spectrometer, multiple sensor channels can be imaged to a two-dimensional CCD array in parallel. Because the wavelength of the signal is transformed in one spatial dimension, different channels can be mapped to different lines of the CCD array.

7.2.2.2   Edge filter-based systems

This design uses a broadband unpolarised light source [9] in order to illuminate the sensor fibre. Light that is reflected by an FBG sensor enters two optical edge filters that overlap spectrally. The two resulting intensities are measured with photo-detectors at the edge filter outputs. Determining the intensity ratio of the photo-currents yields the spectral response of the FBG sensor (Figure 7.3).

Compared to spectrometer and tunable laser approaches, where the bandwidth of the systems is limited by the rates of CCD read-out, respectively, wavelength

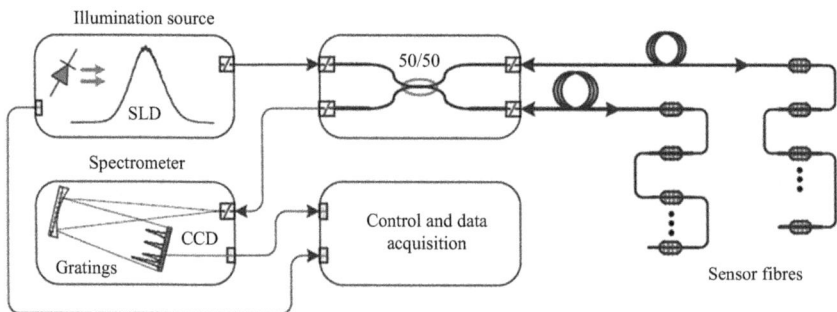

Figure 7.2   Spectrometer-based interrogation unit. A broadband light source illuminates the FBG sensors, the responses are determined by a spectrometer

tuning, this measurement principle shows a higher bandwidth since the sensor response (ratio of two intensities) can be evaluated by means of analog hardware. Thereby, sampling only occurs when the evaluated response is digitised and stored.

7.2.2.3   Tuning laser-based systems

This design uses a narrowband laser as light source that scans through the optical spectrum. Laser pulses at ascending wavelengths are sent to the FBG sensors. The intensities of light pulses that are reflected by the sensors are measured by a photodetector. Since the wavelength of the pulses is known, just the intensity magnitude of the reflected light is of interest. The sensor responses are determined by centroid calculation of the reflected intensities (Figure 7.4).

7.2.3   MG-Y tunable laser technology

The tuning laser, which is the core element of the scanning laser (SL) interrogator, is a monolithic diode laser of which the wavelength can be controlled electronically. Basically two resonators are combined within the structure of this laser diode. Both resonators share the same partially transmissive end mirror at one end. At the other end, a grating structure acts as second reflector as shown in Figure 7.5.

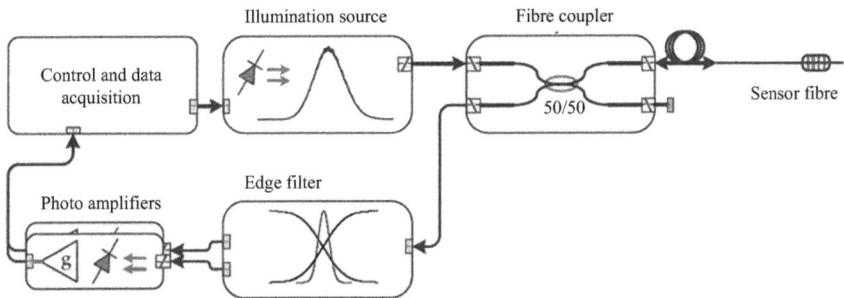

Figure 7.3   Edge filter-based interrogator. A broadband illumination source illuminates the sensors, edge filtering determines the sensor response

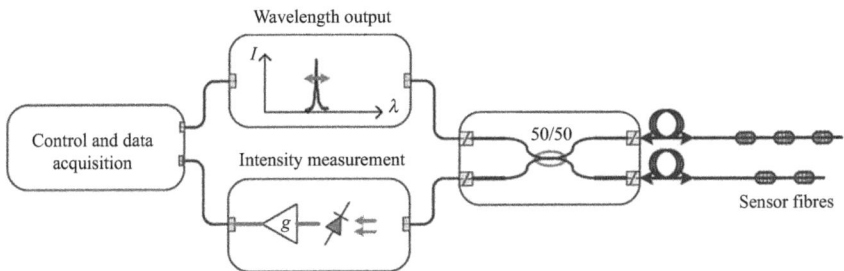

Figure 7.4   Tuning laser-based interrogator. The output wavelength of the laser is tuned, thereby the spectral sensor response is scanned

The modulated-grating Y-structure (MG-Y) laser diode emits light at a wavelength where modes of each resonator overlap, as schematically shown in Figure 7.6. The grating structures can be modulated by carrier injection allowing for RI modulation of the grating material. Electric currents supplied to the grating reflectors adjust the free spectral ranges (FSR) of the resonators which originally are 630 GHz (5.05 nm) and 700 GHz (5.61 nm) [10]. Thereby, the spectra of both resonators can be shifted against each other, which allows to bring different modes of the resonators to overlap. By this, the so-called Vernier effect [11] wavelength tuning is achieved.

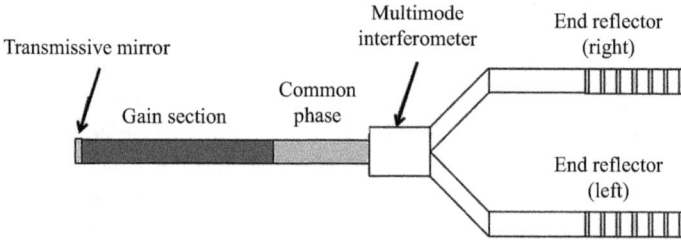

Figure 7.5   *Structure of the modulated-grating laser diode. The resonator is split into two arms, the Vernier effect is used for wavelength generation*

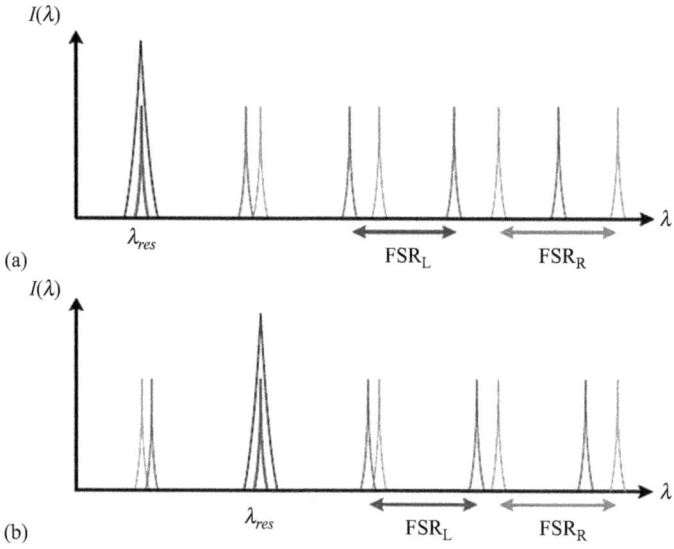

Figure 7.6   *Additive Vernier effect: (a) An output wavelength λ_{res} (or wavelength response) is generated at the spectral position where peaks of left (FSR_L) and right (FSR_R) resonator overlap and (b) by increasing FSR_R, two other modes overlap and a different output wavelength λ_{res} occurs*

Contrary to sweeping laser sources [12] like external cavity diode lasers [13], the Vernier effect does not support continuous tuning. A set of control currents may or may not yield one single-stable output wavelength. Since both resonators show discrete spectra, the output wavelength jumps arbitrary when the control currents are changed. In order to use the MG-Y laser diode for spectral sampling of FBG sensors, it has to be characterised for all combinations of control currents.

7.3 Space engineering systems characteristics

The term space engineering refers to the design, development and manufacturing of systems intended to be used in space. Several boundary conditions have to be taken into account and are therefore outlined in this section.

7.3.1 Design constraints

When a space mission is planned, the requirements for the systems on board the spacecraft have to be determined at first. This is done by comparing the mission parameters (e.g. launcher, platform, orbit altitude and lifetime) with former successful missions. Furthermore, analysis and simulations concerning the new or changed boundary conditions are carried out.

When systems and subsystems are designed and developed, it has to be ensured that all requirements are fulfilled. Therefore, the functionality of the system is tested under relevant environment by means of, e.g., thermal vacuum and radiation tests. Different models of the system (e.g. breadboard, engineering, qualification and flight model) are built for consecutive stages of development [14]. In the case of complex mission scenarios, flight demonstration of critical subsystems is accomplished. The Laser Interferometer Space Antenna (LISA) mission, for example, utilises the technology demonstrator LISA Technology Pathfinder as forerunner mission wherein key technologies shall be verified [15].

The maturity of a system is characterised by the technology readiness level (TRL), a classification of systems and components given by Canadian Space Agency (CSA), ESA and National Aeronautics and Space Administration (NASA) [16,17]. Definition of TRL:

1. Basic principles observed and reported (scientific research).
2. Technology concept and/or application formulated (applied research).
3. Analytical and experimental critical function and/or characteristic proof of concept (laboratory experiment).
4. Validation in laboratory environment (laboratory testing).
5. Validation in relevant environment (environmental testing).
6. System/subsystem model or prototype demonstration in a relevant environment (ground or space).
7. System prototype demonstration in a space environment.
8. Actual system completed and 'flight qualified' through test and demonstration (ground or space).
9. Actual system 'flight proven' through successful mission operation.

A system is space qualified and therefore allowed to be implemented on board the spacecraft after it has been verified that a TRL of eight is achieved. Contrary to the development of ground-based systems, engineering of space-borne systems has to take into account special boundary conditions. Some aspects that have to be considered during development are outlined in the following list.

7.3.1.1   Compatibility with environment

The development of systems that operates on a spacecraft is mainly driven by the environmental conditions that the system is faced with during its lifetime. The environment, a space-borne system has to be designed for, depends on the mission parameters, whereby the altitude mainly determines the in-orbit requirements for the system [18]. A space-borne system has to cope with several different environments that occur sequentially:

- Transportation from development site to launching base: Expected conditions are temperature ranges from $-15°C$ to $+40°C$, humidity, shock and vibrational loads during vehicle-based transfer and possible solar radiation. Usually the fibre-based system is not harmed by this environment since it can be thermally controlled and appropriately packed into protection covers.
- Launch, transportation and in-orbit operation during mission: Radiation, vacuum and microgravity during operation of the system in space and high shock and vibration loads during launch are the main design drivers. The harsh environment in space which depends on the orbit, the kind of spacecraft that accommodates the system and mission specific parameters affect the system during the mission. The system has to be designed and developed such that it withstands the environmental loads.
- Transportation from mission orbit back to earth: Missions where the system is sent back to earth after its mission lifetime, e.g. for data evaluation, are possible but are not elaborated in this chapter.
- Autonomous work: Unlike devices deployed in laboratories, systems working in space must be stand-alone operable. One way is to design the system intrinsically stable so that readjustment is not necessary. If intervention is unavoidable, monitoring and adequate control mechanisms have to be implemented in order to either automatically or by remote control readjust the system.
- Lifetime assessment: It has to be ensured that the entire system with all subsystems and components operates failure tolerant within the expected lifetime. Methods like derating of components or implementation of system redundancy by, e.g., parallel circuiting of electronics can be used [19].
- Power efficiency: Because of the reduced energy disposition, high efficiency and low-power consumption is required. Also, in view of thermal controlling inside the spacecraft, a minimised waste heat is desired.
- Dimensions and weight: The reduced dimension of the satellites requires compact system design to achieve low volume and weight.
- EM compatibility (EMC): All electronic and optoelectronic equipment have to be EMC safe which requires special design [20]. Furthermore, the system itself must not harm surrounding equipment by generating EM radiation.

- Safety margins: In order to ensure correct system operation even if there are deviations between expected and actual operation conditions (e.g. unforeseen solar flares that result in enhanced radiation exposure), safety margins have to be applied to all requirements. Compliance with increased requirements is guaranteed by enhanced qualification testing [21].

7.3.2   Development guidelines

In former days, no uniform system of space standards was available [22]. Therefore different companies used different standards. Divergences between the standards resulted in misunderstandings between companies and thereby in failures of developed systems. Additional effort was necessary to compare the standards among each other which yielded higher development costs. The European Cooperation for Space Standardization (ECSS) was founded in 1993. ECSS is an association of ESA, national space agencies and a consortium of companies from the industrial sector. These partners singed a contract for development of a common framework of standardisation. It aims to create and maintain common standards for space-related development. Thereby, cost efficiency, performance enhancement and competitiveness of the European space industry on the world market shall be achieved and enhanced.

The policy of ECSS was defined in one separate document and arranged the principles of the new system. Key points of ECSS policy are as follows:

- ECSS standards have to be made applicable to a project by a contracting party; they are not legal by themselves. The contracting party that introduces an ECSS standard as requirement document is responsible for control of compliance of the development with the standard.
- Requirements shall be defined target oriented rather than task oriented: The performance that has to be achieved rather than the description of the way to achieve it is the objective of a standard. This leaves the way open for new technological developments.
- ECSS standards shall not duplicate other norms, e.g. International Organization for Standardization or European Norms standards. Instead links to already existing and complementary standards shall be included.

From the very beginning of ECSS, this standardisation was intended to stay in motion since it is continuously adapted to changes in technology.

7.3.3   Space radiation

Radiation is one of the major concerns for systems operating in space. Properties of electronic and optic equipment are altered when irradiated with particles (e.g. protons, electrons, heavy ions) or with EM waves (e.g. X- or gamma rays, UV light). Even system failure due to radiation occurred in former missions. As a matter of fact, a latch-up damage of a random access memory chip on board the precision range and range-rate equipment satellite due to proton impact, for example, leads to the loss of the satellite 5 days after launch [23]. Implementation of a latch-up detection circuit during development and controlled reset of the system may have resulted in a failure-tolerant operation of the satellite.

The Earth is subjected to a nearly isotropic flux of charged particles, beta and gamma rays. The radiation flux is mainly generated not only by solar flares of the sun but also by omnidirectional galactic cosmic radiation originated outside the solar system. The flux of the charged particles consists of about 85% protons (hydrogen nuclei), 14% alpha particles (helium nuclei) and 1% heavier ions (e.g. iron and carbon nuclei) [24].

A particle is only able to pass the magnetic field of Earth by having a certain energy level (below 1 GeV at the poles, up to 17 GeV at the equator). Incoming particles of lower energy are deflected toroidal round the earth by Lorentz forces. Research lead by James Van Allan revealed that two radiation belts with high concentration of trapped particles exist around Earth. The inner belt at an altitude around 3,000 km mainly consists of protons and electrons at fluxes of 2×10^5 particles/cm^2 s and 3×10^6 particles/cm^2 s for energies above 10 and 1 MeV, respectively. The outer belt at an altitude of approximately 25,000 km consists of electrons at a flux of 2×10^6 particles/cm^2 s (energy >1 MeV) [25,26].

7.3.4 Radiation effects

When matter is exposed to radiation, the effects of interaction depend on several parameters: mass, charge, kinetic energy, incidence angle of the particle or ray, type and density of the targeted matter. Degradation that occurs in electronic or optical components is due to either ionising or atomic displacement (AD) effects dependent on the type of particle or ray:

- Ionisation: Charged particles interact primarily by Rutherford scattering (Coulomb scattering). This interaction can cause excitation or ionisation of atomic electrons (electron-hole pair generation) [27]. Total-ionising dose (TID) is the measure of ionising radiation. The unit of TID is rad or Gy, whereby 1 Gy = 100 rad.
- AD: If sufficient energy is transferred to atoms, they are displaced from their normal lattice positions. Heavy-charged particles can cause elastic or inelastic scattering. In an elastic collision, the bombarding particle transfers a portion of its energy to an atom of the target material and can dislodge the atom from its lattice position [28]. Non-ionising energy loss (NIEL) is the measure of non-ionising radiation effects. The unit of NIEL is MeV/g. Charged particles, e.g. heavy ions, which incident on matter can cause both ionising and displacement effects.
- Radiation effects on semiconductors: Ionising radiation results in cumulative energy deposited in a given volume. Most of the converted energy of an incident particle creates electron-hole pairs. At this ionisation, the valence band electrons in the solid are excited to the conductor band, thereby enabling *n*-type conduction in the conductor and *p*-type in the valence band. This produces a variety of device effects like energy band shifts or leakage currents which can, for example, yield gate threshold voltage shifts in complementary metal oxide semiconductor devices [29]. Furthermore, high energies of radiation lead to internal charging of conducting layers in electronic devices. If high potential

differences between layers separated by insulators are reached, electrical breakdowns and damage are caused. Sources that ionise the material are electrons, protons and Bremsstrahlung, which is a secondary radiation due to slowdown of incoming particles like heavy ions.

Ionising-based damage of electronic devices shows the following characteristics [29–31]:

- Biasing of semiconductor devices leads to higher radiation damage compared to unbiased operation.
- Trapped charge reduces after irradiation dependent on the temperature and applied electric field, whereby the damage is partially or completely cured.
- Although some devices are no longer sensitive to dose-rate effects below approximately 1 rad(Si)/s, other devices continue to be affected even at dose rates of 0.002–0.005 rad(Si)/s [32]. In linear devices based on junction-isolated bipolar transistors, the enhanced sensitivity of the "low-dose rate" characterizing this type of devices will produce higher damage [32–35].

AD is caused by incoming high mass particles such as protons and heavy ions. Atoms are thereby displaced from their crystal lattice position leading to stable defects created within the bandgap. AD is of major concern for bipolar transistors because these devices are based on minority carrier conduction [36].

7.3.4.1 Radiation effects on optical components

Physical effects, that occur when optical components are irradiated, are not fully understood [37]. During irradiation of optical components, fluorescence can occur and influence surrounding equipment [38]. Long-term degradations in transparent materials are transmission losses. These losses are due to the presence of crystalline defects, so-called colour or F-centres (from the German word 'Farbe') [39,40]. In particular, irradiation with heavy particles alters the lattice structure, exposing optical material to radiation results in the displacement of lattice ions and thereby in the creation of lattice vacancies. If an electron from the valence band is excited into this vacancy, the band gets trapped and an F-centre is created within the bandgap of the optical material between the valence and conduction bands [41]. Electrons in such a vacancy absorb photons in the visible spectrum such that the transparent optical material becomes coloured. If occurring radiation, e.g. light, has a very high energy or intensity, the electron may be released from the vacancy. Thereby, the material gets (photo-)bleached and loses its colour.

7.3.4.2 Radiation damage in optical fibres

At NASA, databases of radiation test results on several types and brands of optical fibres are available [42,43]. In general, optical fibres have been used for space applications for more than 30 years [44] without failures. Rare earth-doped optical fibres however are well known to be highly radiation sensitive [45–47]. In other areas of research, e.g. for nuclear power plants, the radiation sensitivity of rare earth-doped fibres is used for dosimetry [48–50].

Guiding high intense light through attenuated fibres allows to cure the defects by photobleaching [51].

Preliminary examinations performed by the Max Planck Institute of Quantum Optics (Germany) in cooperation with Kayser-Threde GmbH (Germany) evaluated all types of fibres implemented in an erbium-doped optical frequency comb [52]. Several tests with different radiation sources have been conducted whereby the transmission loss at the wavelength 1,310 nm was measured. Standard telecommunication single-mode fibres and highly non-linear fibres showed transmission losses of less than 0.01 dB/m after irradiation with a total dose of 955 Gy. The erbium-doped fibre showed transmission losses of 3 dB/m after a total dose of only 190 Gy and 10 dB/m after 380 Gy. During the irradiation with 20 MeV protons, a ytterbium-doped fibre was tested additionally. After irradiation of both rare earth-doped fibres with 12×10^{13} protons/cm^2, the transmitted intensities were reduced by 1.55 dB for the erbium and 0.65 dB for the ytterbium-doped fibre for lengths used in fibre combs.

Radiation exposure causes F-centres which degrade the performance of the fibres by attenuation. Several radiation test results on rare earth-doped optical fibres are available in literature [53–55], but test conditions like dose rates, total dose and dopants of the fibre vary for each experiment. Ytterbium-doped fibres are known to be less sensitive to radiation than the erbium-doped ones [56].

Up to now, it has not been clarified, which doping composition is best suited for low radiation sensitivity, but hydrogen loading is a technique for radiation hardening [57–59]. Fibres are therefore coated with metal or carbon and loaded with hydrogen under high temperature and pressure. The coating prevents the hydrogen from out-gassing. It has been shown by the Fibre Optic Research Center (Russia) that hydrogen loading of hermetically coated erbium-doped fibres extends their lifetime in space by more than five times. Exposing the fibres to Gamma radiation from a cobalt-60 source (0.028 Gy/s) to an accumulated TID of 2 kGy only reduced the lasing efficiency at 980 nm by 1 dB. Pumping the fibre at a wavelength of 980 nm enhances photobleaching effects which supports the radiation hardness of the fibre [60].

7.3.4.3   Radiation effects on fibre Bragg gratings

Irradiation of FBG in optical fibres results in a shift of the spectral response of the sensor due to a change of the RI $n1$, $n2$ and $n3$ of the fibre. Spectral shifts around 25 pm for total doses up to several hundred krad have been observed [61,62]. Recovery of the FBG spectral shifts is possible after several hours [63] of annealing or by applying even higher doses of several Mrad [64]. Writing FBG by femtosecond-pulsed UV lasers makes the spectral shift of the sensor independent from the fibre material [65]. Width and amplitude of the FBG spectrum are not altered by radiation. Furthermore the fibre coating has an influence on the wavelength shift, when the fibre is stressed by, e.g., shrinking of the coating due to radiation [66]. Long-term irradiation of FBG sensors over a period of 40 months in a nuclear reactor revealed suitability of FBG sensors [67]. The selection of the fibre material and the fabrication technique of the FBG enable to produce sensors that are less sensitive to radiation [68]. Contrary to rare earth-doped fibres, where hydrogen loading decreases radiation sensitivity, FBG sensors should be written in photosensitive fibres with a high germanium concentration [69].

7.3.5 Simulation of radiation exposure for orbital missions

In order to design and develop a satellite system, the expected radiation environment of the planned mission has to be identified with high care. This can be achieved through simulation software, e.g. Space Environment Information System (SPENVIS) provided by ESA [70].

The Earth orbit is usually divided into three regions: LEO, medium Earth orbit (MEO) and geostationary Earth orbit (GEO). The simulation parameters given are used for achieving the results shown in Figures 7.7 and 7.8, respectively.

Various simulators can be used to determine the accumulated dose of EM and particle radiation for user defined orbits in dependence of shielding. Figures 7.7

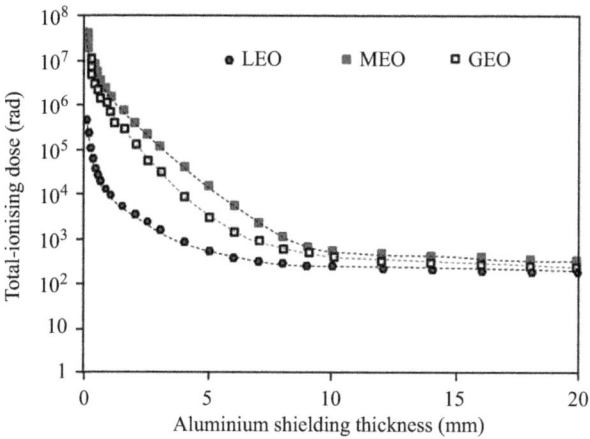

Figure 7.7 Accumulated TID within a mission duration of 1 year according to the orbit parameters given in Table 7.1

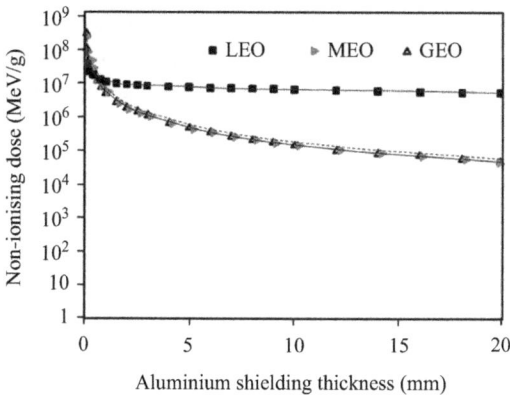

Figure 7.8 Accumulated non-ionising dose within a mission duration of 1 year according to the orbit parameters given in Table 7.1

and 7.8 show the simulation results for the three main orbit types LEO, MEO and GEO according to Table 7.1.

The TID (Figure 7.1) in MEO is higher than in GEO due to the outer Van Allan radiation belt. Therefore, most of MEO's total dose is due to electron irradiation. Because of the space inner belt that consists mainly of protons, the non-ionising dose is highest for LEO. A radiation sensor on board the Galileo In-Orbit Ver-ification (GIOVE) satellite verified SPENVIS simulations carried out prior to the GIOVE mission, although with divergences within measurement accuracy [71].

7.3.5.1   Microgravity and vacuum

Thermal effects on space-borne systems occur due to microgravity and due to vacuum. During lift, earth's atmosphere is left whereby the outside pressure of the spacecraft reduces. If gas inside the system has to be released, some kinds of vents are necessary to ensure that the housing of the system is not harmed by forces due to remaining pressure differences.

7.3.5.2   Thermal issues

Due to vacuum and the absence of gravitation, thermal convection is not available for heat transfer (unless pressurised housings and fans are used). Therefore, only thermal radiation and thermal conduction can be used for waste heat dissipation.

Thermal radiation only works if the surrounding equipment serves as heat sink at a lower temperature than the system that dissipates heat. If this is not the case in a spacecraft, copper bands connect the heat source inside the spacecraft with a heat sink that is located outside and points to deep space. The design of the heat con-ductor takes into account the required operation temperature and amount of waste heat of the system. Adequate design of the thermal interface can be challenging, if the system has different operation modes that produce different amounts of waste heat. Thereby the heat conductor has to be designed such that the system keeps its operating temperature within the range from minimum to maximum produced waste heat. If this is not possible either an active temperature control, e.g. heat pipes, or electric heaters have to be used. Another approach can make use of a pressurised housing where the system is encapsulated. Due to the absence of gravity, ventilators have to be used to enable heat exchange by thermal convection inside the housing. This solution is only possible, if the housing is colder than the heat source.

Table 7.1   Earth orbits characteristics

Earth orbit	Low	Medium	Geostationary
Altitude (km)	100–200	2,000–35,000	35,786
Typical mission	Earth observation	Global navigation satellite system: GNSS	Telecom
Simulated parameters:			
Altitude (km):	820	23,600	36,000
Inclination (°)	98.0	56	0

Temperature control and dissipation of waste heat by conduction result in enhanced complexity of the system. This goes along with higher development and production costs and reduced overall efficiency of the system. Therefore, the development of space-borne systems aspires to an enhanced temperature range the system can cope with, or to internal heat exchange mechanisms whereby external heat transport is minimised.

7.3.5.3 Outgassing issues

Synthetic materials, e.g. polymers or lubricants, can include plasticisers that diffuse out of the material under vacuum conditions. One negative effect thereof is reduced flexibility or altered characteristics of the material. Moreover, the freed molecules can cause problems if they contaminate surfaces from, e.g. free space optical components. Outgassing strongly depends on temperature: the higher the temperature, the faster is the outgassing progresses. Thus, procedures like thermal cycling or out-baking of materials can be carried out to intentionally cause outgassing of materials prior to their use as components in the system. Outgassing is attended by mass loss due to the freed molecules. ESA and NASA hold databases with qualified materials that show a tolerable total mass loss (TML) [72]. The lower the TML, the better the material is suited for space applications.

7.3.5.4 Shock and vibration

High vibrational loads with accelerations up to several hundred g dependent on the launcher that is used (e.g. Ariane, Sojus or Vega) affect the system during launch. Usually the launcher has several stages that are ignited consecutively.

Every time one stage is burnt-out, it is separated from the launcher by explosive charges. Thereby, shock loads up to several thousand g appear. Within the user manual of the Ariane 5 launcher [73], shock values up to several thousand g are given as worst case assumption. This has to be taken into account, when a system is design for a mission that is launched by Ariane. The shock and vibration loads that penetrate the instrument inside the satellite housing are lower due to damping of spacecraft and instrument mountings. If, however, a sensitive or alignment critical part of an instrument has to cope with such requirements, either the mounting has to be mechanically robust enough or additional locking mechanisms have to be implemented.

Special devices are available that lock, e.g., moving parts of a system to a save position during launch. As soon as the satellite is in orbit, the mechanism releases and normal operation can start. The challenge is not to construct a locking mechanism that withstands the shock load but to make sure that it is able to release by all means even after high loads have been applied to the mechanism.

7.4 Development of optical fibre sensing systems for space

Within this section, the development of OFS for applications in space is summarised. A trade-off between different interrogation principles is carried out.

The interrogation principle of the selected SL technology is also described. A typical example of the architecture of the interrogator hardware is illustrated and the characterisation of the SL interrogator which needs to be performed prior to the operation of the system is also explained.

7.4.1   Optical fibre sensing interrogation technology for space applications

OFS is on the way towards its utilisation for space applications. As mentioned and detailed before, the main advantages of optical sensors like FBG desired in space applications are lightweight, insensitivity to EM disturbances, ease of distribution and scalability are natural properties of OFS and requirements to a sensor system of a spacecraft. The possibility to implement several FBG sensors in one single-sensor fibre without influencing each other and to install such fibres in composite materials during their fabrication is also desirable for space systems.

Launchers (e.g. Ariane) and satellites are currently monitored by hundreds of electric sensors during test and qualification. OFS is regarded as a potential technique to overcome limitations of recent electronic-based monitoring systems. Instruments based on tunable lasers are established devices for demodulation of FOS. These state-of-the-art systems built on tunable laser sources often use free beam set-ups for wavelength tuning. Thus, they are sensitive to environmental influences like vibration or temperature impacts. In addition, by using fibre-coupled photodetectors for input intensity measurement, none of the subcomponents is sensitive to environmental disturbances, provided that the laser diode is adequately temperature controlled.

Regarding their implementation on board spacecraft, the interrogation techniques summarised in Table 7.2 show different advantages and drawbacks.

Table 7.2   Trade-off between OFS interrogation technologies. The majority of requirements for sensing on board a spacecraft are fulfilled by the scanning laser interrogation technology

Requirement	Spectrometer interrogator	Edge filter interrogator	Scanning laser interrogator
Technology maturity	High	Medium	Medium
Robustness	Medium	High	High
Compactness	Low	Medium	High
Service-free operating time	High	High	High
Operational lifetime	High	High	High
Power efficiency	Medium	Medium	High
Measurement performance (accuracy and rate)	Low	High (suitable for strain sensing)	Medium
Number of sensors/complexity of the system	Medium	Low	High

7.4.2   Fibre Bragg grating sensor-based scanning laser interrogation principle

The SL interrogator is based on a tunable laser that is capable of providing discrete wavelengths within its available spectrum, e.g. from 1,528 to 1,571 nm. The hardware architecture of the SL interrogator results in a very flexible and adaptable system. The main advantage is the wavelength-switching capability compared to standard interrogators of the same measurement principle [74]. Since every wavelength within the output spectrum of the MG-Y laser diode is accessible within the same switching time on microsecond timescale, spectral areas without information can be omitted during standard operation, and each sensor can be measured individually. Spectral gaps between two consecutive sensors can thereby be skipped and overall measurement rate is increased.

In order to detect the peak wavelength of FBG sensors, the SL is able to scan through a spectrum in the infrared (IR) region with a width of more than 40 nm. A sensor array can be connected to one measurement channel by implementing several sensors with different design wavelengths inside one sensor fibre. The SL interrogator searches for the sensor responses and is then able to interrogate user-chosen sensors sequentially. The magnitudes of the reflected intensities depend on the actual sensor position that is determined by the measurand (e.g. temperature). One single sensor is scanned by a variable number of spectral sampling points and the spectral answer of the sensor is then calculated by centroid algorithms. Depending on the spectral width of one sensor, the number of sensors that shall be interrogated and the required sampling points per sensor (Figure 7.9), a maximum sampling frequency of 10 kHz is achievable with the SL interrogator.

In order to determine the current spectral position of an FBG sensor, the spectrum is scanned with laser pulses generated by the MG-Y laser diode. Therefore the SL generates a sequence of output laser pulses according to the digital-to-analog

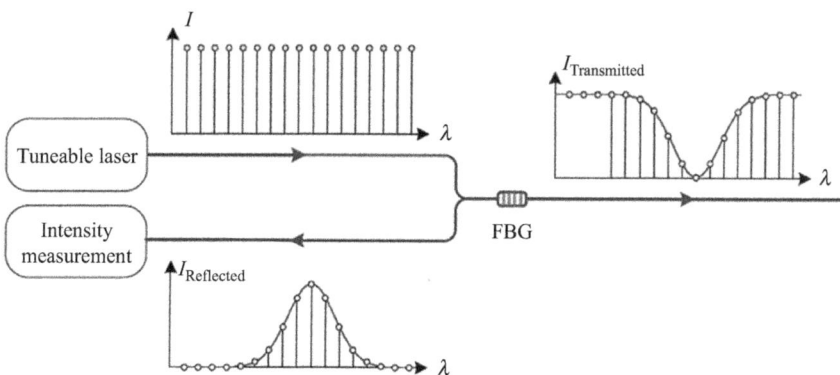

Figure 7.9   The SL system emits laser pulses at ascending wavelengths. Pulses of which the output wavelengths match the design wavelength of the FBG sensor are partially reflected. The sensor response is determined by measurement of the reflected intensities

(DAC) values stored in the lookup table. This is resulting in a train of laser pulses (Figure 7.10(a)), wherein succeeding pulses show rising wavelengths. Actually, the laser emits light continuously and not as pulses, but in terms of laser wavelengths a pulse train occurs. When the wavelength matches the design wavelength of an FBG sensor, the emitted light is partially reflected by the FBG and guided back to the photodetector of the corresponding channel. The field programmable gate array (FPGA) that evaluates these input intensities compares measured intensities with a user defined noise floor level in order to distinguish between sensor responses ($I_{in} > I_{nf}$) and noise coming from background light or from the electronics of the photoamplifiers ($I_{in} < I_{nf}$).

Figure 7.10(b) shows a typical sensor response. The two intensities lower than the noise floors at the edges of the sensor response define the spectrum of the sensor peak. The wavelengths corresponding to these two intensities are entitled 'λ_{start} (λ_s)' and 'λ_{end} (λ_e)' as they are used as start and end point for sensor sampling. The respective sensor response in the wavelength domain is shown in Figure 7.10(c). Since the wavelengths are non-equidistantly distributed in the spectrum, the number of sample points per sensors may vary, when the sensor shifts spectrally due to measurand changes.

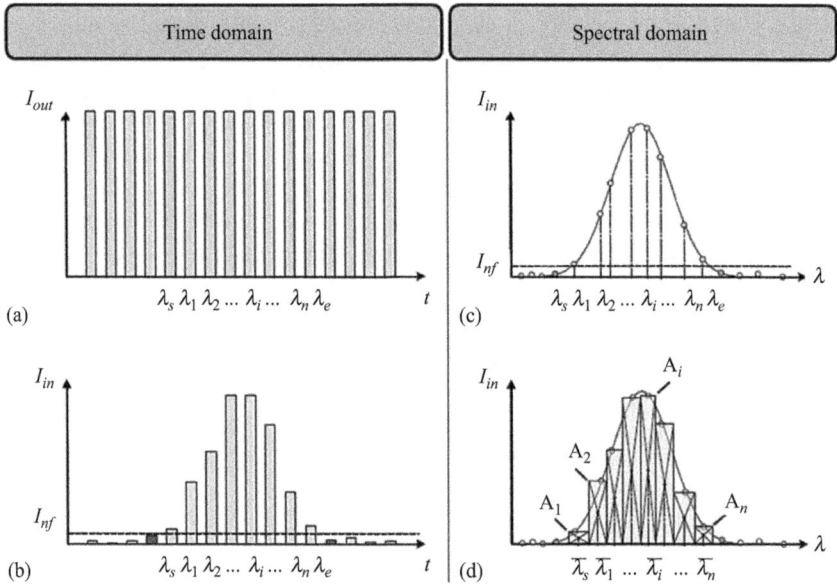

Figure 7.10   Time and wavelength domain of one FBG sampled by a variable number of sampling wavelengths. (a) Train of laser pulses, (b) typical sensor response, (c) sensor response in the wavelength domain and (d) sensor response after noise subtraction. Due to a measure and change, the sensor answer is spectrally moving to higher or lower wavelengths. The special centroid algorithm determines the mean wavelength of the sensor response

7.5 Experimental data acquisition fibre optic sensor demonstrator on the Europeans Space Agency's' PROBA-2 Satellite

Spacecraft monitoring is vital for successful operation during all mission phases; ground qualification, launch and mission flight operations [1]. A distributed network of various sensors is needed to provide information on the spacecraft health and performance of subsystems and payloads, and particularly to estimate the amount of available propellant [75–77]. The monitoring requirements include the volume, pressure, leakage detection, temperature distributions, valve/actuator status. Currently an *ad hoc* array of electronic sensors is used requiring individual-shielded wiring and signal processing that adds mass and limits the number of sensors and their accuracy. The electronic sensors, while well established, have a number of disadvantages in the operational environment of a spacecraft, including

- susceptibility to EMI with costly and complex signal routing,
- susceptibility to sparking and the resultant explosion risk,
- poor direct chemical compatibility with propellants such as hydrazine,
- poor measurement response times and reduced measurement accuracy due to the required shielding and mechanical-safety barriers,
- limited sensor multiplexing with typically one sensor per twisted wire pair, resulting in large signal harnesses,
- a significant mass penalty due to shielding requirements for the electrical data harness and
- the corresponding signal-processing electronics typically have to be located close to the sensors, exposing them to the harsh and noisy environment of the rocket propulsion system.

Table 7.3 summarises the minimum ESA performance requirements for the various sensors employed for the rocket propulsion subsystem.

One of the first solutions to the monitoring requirements of spacecraft based on laser-based sensors was in the fibre sensor demonstrator (FSD) flown on ESA's PROBA-2 launched in 2009. This was the first achievement of a full fibre optic network mounted on a satellite so far. By combining serial wavelength-division sensor multiplexing and parallel signal distribution along single strands of fibre, we were able to meet n a high sensor capacity. Our FSD is EMI insensitive, allow a remotely positioning of the interrogation system, in addition to offer a flexible signal routing using lightweight fibre optic cables with microtubing armour for strength and picometer range measurement resolution of the sensor spectral response.

This section discusses the preparation and ground testing of the FSD for PROBA-2. The system uses a novel generation of a tunable fibre laser source functioning at 1,500 nm for both the illumination and interrogation of multiple, parallel strands of WDM-coupler-based FBG or Fabry–Pérot optical sensors. The interrogation system can be employed efficiently with a wide variety of sensor types including pressure, strain, temperature, valve status and gas leakage.

Table 7.3   Standard methods of propellant gauging technologies

Sensor propulsion	Subsystem unit	Range	Accuracy	Current technology
Pressure transducer (low)	Propellant tank, cold-gas feed system	0.1–3 MPa	0.01%–0.1% (full scale output (FSO))	Strain gauge, quartz crystal
Pressure transducer (high)	Pressurant tank	>20 MPa	0.1%–1% FSO	Strain gauge, quartz crystal
Temperature (low)	Propellant	Tank, pressurant tank	−20.15°C to +49.85°C	1%–2% FSO
Temperature (high)	Thrust chambers	−20.15°C to 2,226.85°C	1%–2% FSO	Thermistors, thermocouples
Position indication	Valves	1,000,000 cycles in satellite lifetime	Reliability >0.9999	Reed switch, microswitch
Flow measurements	Propellant, gas	0.5–500 g/s	<0.5%	Turbine, ultrasonic
Volume measurements	Propellant, tank	<200 l end-of-life fuel	<0.5%	Pressure temperature
Leakage measurements	Propellant	Gas 10^{-6}–10^{-3} sccm	Factor of 10 in full scale	Spectrometry

The fibre optic interrogation unit provides a high-output spectral power density that can provide sufficient optical power to probe several hundred FOS with a high signal-to-noise ratio (SNR) ($>10^5$). A special architecture is used to provide redundancy of the critical components. Sensor parallel and serial multiplexing architectures facilitate accommodation of hundreds of sensors with a single-fibre laser interrogation system. The current measurement repetition rate is as fast as 0.1 s to facilitate transient measurements.

Following the FSD success, several novel FOS have been developed employing hybrid microelectromechanical systems (MEMS)/FBG structures that enable different types of measurements (pressure, temperature, stress, valve status and flow) using a common fabrication FBG methodology. The significant advantage of that concept is that different FOS can use the same interrogation system and can be serially WDM coupler on single strands of optical fibre. Examples of the sensors developed include the following:

- Novel combined *P/T* sensor for high-reliability *in situ* measurement of the pressure of propellant tanks (0.01% of full scale).
- High-temperature sensor for thruster monitoring to above 400°C.
- MEMS-enhanced FBG temperature sensors for high-accuracy temperature measurements (0.05% of full scale).

To meet the requirements of the space environment, a proprietary lightweight, flexible cabling was developed that is only about 0.9 mm outer diameter (OD) yet provides hermetic sealing for the single-mode silica core, vibration dampening and crush-resistant mechanical protection. The resultant sensor signal distribution harness is lightweight (i.e. below 0.3 g/m) and provides high signal integrity with EMI immunity and flexible routing with a minimum bend radius of about 15 mm.

The assessment of the FSD FOS technologies included consideration of chemical compatibility with propellants, such as hydrazine, the space radiation environment (UV, gamma rays, atomic oxygen, protons), vacuum thermal cycling (−40°C to +60°C) and susceptibility to the mechanical environment (random vibrations, launch shock).

7.5.1 Fibre optic sensor demonstrator system

A schematic of the FSD system is shown in Figure 7.11. The system features a central interrogation system weighing below 1.2 kg and requiring less than 3.5 W peak power (Figure 7.12(a)). FSD's interrogator is 12×70×150 mm in size. The interrogator is located remotely from the fibre sensors, at an opportune location in the PROBA-2 spacecraft to optimise the usage of available space. The FSD electronics feature two electronic printed circuit boards (PCBs) integrated on an aluminium support frame. The central processing unit (CPU) PCB features a fault-tolerant architecture with resettable latch-up protection, an FPGA microprocessor and 768 kB of static random access memory (SRAM) for the data storage. Dual electrically erasable programmable read-only memory enable updating the software during the flight. The FPGA processor will provide real-time analysis of the FBG sensor peak positions to significantly compress the FSD measurement raw data size for ground downlinks.

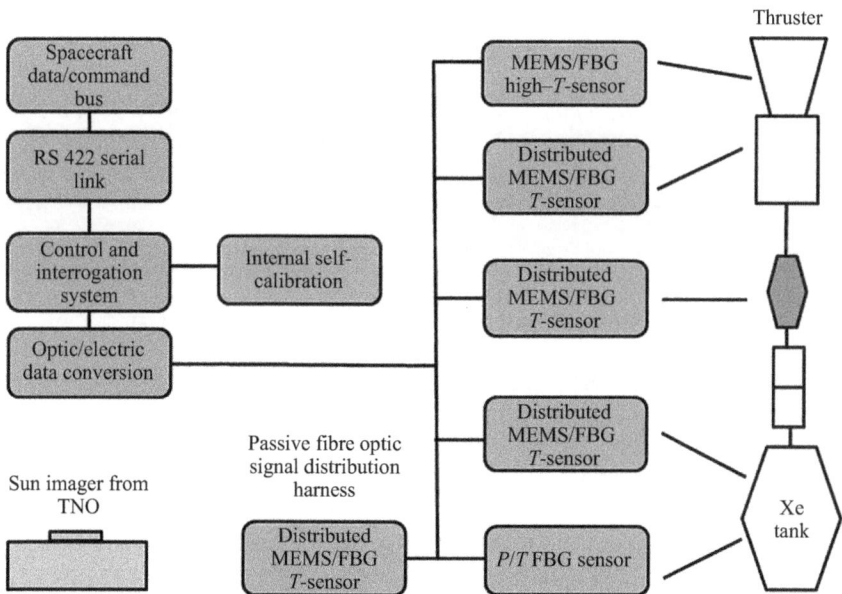

Figure 7.11   Block diagram of the FOS demonstrator for PROBA-2

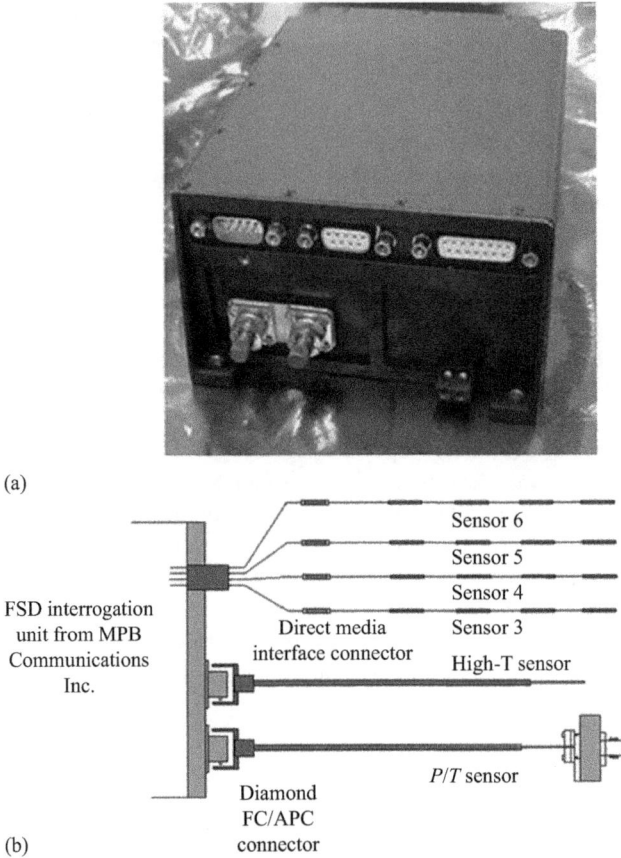

Figure 7.12   (a) Photograph of FSD rep unit interrogation system (1.2 kg, 3.5 W) and (b) schematic of the FSD fibre optic signal harness. © 2010 MPB Communications Inc. (Reproduced, with permission, from MPB Communications Inc., Pointe-Claire, Canada)

The FSD fibre optic signal harness consists of six parallel output/return channels:

• One fibre optic cable containing a dual FBG fibre optic *P/T* sensor for the xenon tank pressure (0–45 Bar) with a demountable Diamond® FC/APC connector and 0.9 mm OD armoured jacket.

• One fibre optic cable containing a high-temperature FBG sensor for measurement of the thruster temperature (−40°C to 350°C peak temperature) with a demountable Diamond FC/APC connector and 0.9 mm OD armoured jacket.

• Two strands of WDM-coupler-based temperature sensors, each with an internal temperature-stabilised reference FBG and four external FBG temperature sensors with MEMS mechanically amplified temperature sensitivity for −40°C to 70°C to monitor the temperature distribution at critical points along the Surrey Satellite Technology Ltd (SSTL) propulsion system.

- Two strands of WDM-coupler-based temperature sensors, each with an internal temperature-stabilised reference FBG and four external FBG temperature sensors with MEMS mechanically amplified temperature sensitivity for −40°C to 70°C to monitor the temperature distribution at additional points within the PROBA-2 subsystems.

The selection of the fibre optic cabling is a critical issue due to the impact of the space environment on the optical fibre performance [78]. Fibre optic cabling issues include the following:

- Selection of vacuum compatible materials suitable for use at temperature of 350°C.
- Mechanical protection for sections of fibre optic cable in free space between mechanical attachment points (based on initial vibration testing for the FSD).
- Prevent fibre kinking during strong mechanical vibrations.
- Dampen mechanical impact on silica core during vibrations.
- Hermetic sealing of silica to minimise cracking due to stress corrosion effects associated with H_2O and OH.
- Operation with repeated thermal cycling between −40°C and +70°C.
- Lightweight, flexible cabling.

For the FSD external fibre optic cabling, a copper-coated single-mode silica optical fibre was selected. The copper coating thickness was typically about 20 μm over the 125 μm cladding diameter of the fibre, resulting in a net OD of about 165 μm. The copper-coated optical fibres exhibit much better hermetic sealing against moisture and superior tensile strength than standard telecom optical fibres. Moreover, they are usable to higher temperatures exceeding 500°C for harsh environment applications (melting point of copper is about 1,080°C). The typical minimum bend radius for the single-mode silica fibre is about 6 mm. The proprietary cabling jacket includes additional layers for vibration dampening and mechanical strength to protect the copper cladding and the fibre against abrasion and impact during mechanical vibrations and to prevent kinking during cable routing and attachment. The resultant cable has an OD of only 0.9 mm with a recommended minimum bend radius of 15–20 mm and a linear mass of 0.3 g/m.

Preliminary testing of the FSD fibre optic cabling was conducted and included continuous vacuum thermal cycling between −20°C and 80°C for several days corresponding to about 100 cycles and random vibration between 20 and 2,000 Hz to the PROBA-2 qualification levels.

To enable the control of the fibre optic system and the acquisition of the resultant sensor data, a low-power, fault-tolerant CPU PCB and (data acquisition DAQ) PCB were designed. Figure 7.13(a) shows a photograph of the flight CPU PCB. This features a space grade direct current (DC)–DC converter from Interpoint® and FPGA from Xilinx, Inc., as well as the Xiphos latch-up protection for the DC power lines. Data and command I/O is provided using dual, redundant RS422 differential receivers and drivers. The FSD CPU PCB also contains 768 kB of SRAM, sufficient to store extended high-speed transient measurements during satellite manoeuvres and the PROBA-2 thruster firings. The FSD will operate as a slave to the PROBA-2 processor for the demonstration flight; powering on and measuring the sensor network and downloading the resultant data upon request from PROBA-2. Repetitive measurements can be performed at a selected time interval and initiated at a specified time.

Figure 7.13(a) shows the flight DAQ PCB. This is employing space grade and 883 milgrade components for the critical functions, including a 16 bit analog-to-digital converter and serial interface digital-to-analog drivers. To minimise the power consumption, micropower operational amplifiers (op amps) are employed for the analog signal processing. The DAQ PCB includes the control/monitoring electronics for the tuneable fibre laser diode pumps and filters, as well as six fibre-coupled indium gallium arsenide detector channels with individual preamplifiers.

Figure 7.14 shows the measured spectral scan of an optical fibre containing two FBG written at different wavelengths. The scan was measured at a scan rate of

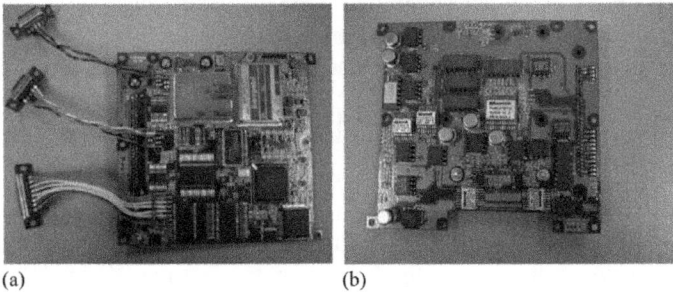

(a)   (b)

Figure 7.13   (a) Photograph of flight CPU PCB and (b) DAQ PCB

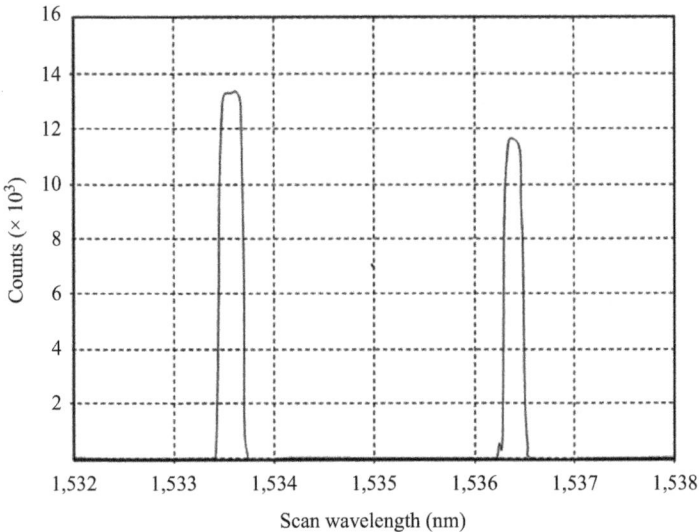

Figure 7.14   Experimentally measured scan of a fibre optic line containing two FBG at different operating wavelengths as obtained using the FSD interrogation unit

0.25 ms per point using the FSD DAQ system and stored in the SRAM. The data was then downloaded *via* the FSD RS-422 interface. With the FSD interrogation unit, operating between 1,520 and 1,565 nm, one obtains the full spectral scan of the grating characteristics with a spectral resolution (1–2 pm) that is comparable to an optical wave meter but with a much higher SNR. The measurement precision, despite the FSD unit compact size and low power, is significantly superior to that of a typical commercial optical spectrum analyser (about 10 pm spectral resolution).

7.5.2 Fibre optic sensor demonstrator sensors

One of the challenges with using FBG for temperature measurements is that the grating centre wavelength (CWL) is sensitive to both temperature and strain. A special proprietary packaging was developed that nearly triples the effective sensor sensitivity to temperature ($\Delta\lambda/\Delta T \approx 0.03$ nm/°C), translating the variation of the wavelength $\Delta\lambda$ with respect to the variation of the temperature ΔT, while decoupling the FBG from the sensor mounting and surface strain. This enables robust mounting of the sensor to the space structure with good thermal contact while maintaining the FBG sensor calibration.

The experimental measurement results are summarised in Figure 7.15(c). Relative to the characteristics of the 'as-prepared' FBG, the new sensor packaging and mounting methodology resulted in nearly identical measured wavelength $\lambda(T)$ FBG CWL characteristics in vacuum at 10 Torr, even after 4 days of continuous vacuum thermal cycling between −20°C and +80°C. As indicated by the experimental measurements, the overall T-sensor mounting and calibration characteristic was very stable in the vacuum environment after extended thermal vacuum cycling.

The FSD innovative P/T sensor uses multiple FBG with special mounting to provide simultaneous pressure and temperature measurements. The pressure readings are independent of the gas composition. The integral temperature measurements are employed to correct for the effects of temperature on the pressure readings. The P/T sensor employs a heat-treated, orbitally welded stainless steel (SS) housing that is suitable for direct contact with propellants such as hydrazine. It can be employed for either single ended or differential pressure measurements. The P/T sensor was proof tested in N_2 at 1,200 psi, relative to the maximum operating pressure of 600 psi. The P-sensor response to the applied pressure is very linear with minimal hysteresis comparable to the reference pressure measurement accuracy (Figure 7.16(b)). The temperature-compensated maximum hysteresis was about ±0.015 nm relative to a full scale of about 12.75 nm for 600 psi. The T-sensor response is totally isolated from the pressure effects.

The FBG fabrication for the high T-sensor entailed special writing and processing steps to provide the sensor stability at high operating temperatures. Figure 7.17(b) shows an assembled high T-sensor with cabling and FC/APC optical connector.

The high T-sensor testing consisted of

- thermal cycling in air between room temperature (RT) and 400°C and
- repeated sinusoidal vibration testing between 2 and 2,000 Hz using a small shaker.

(a)   (b)

(c)

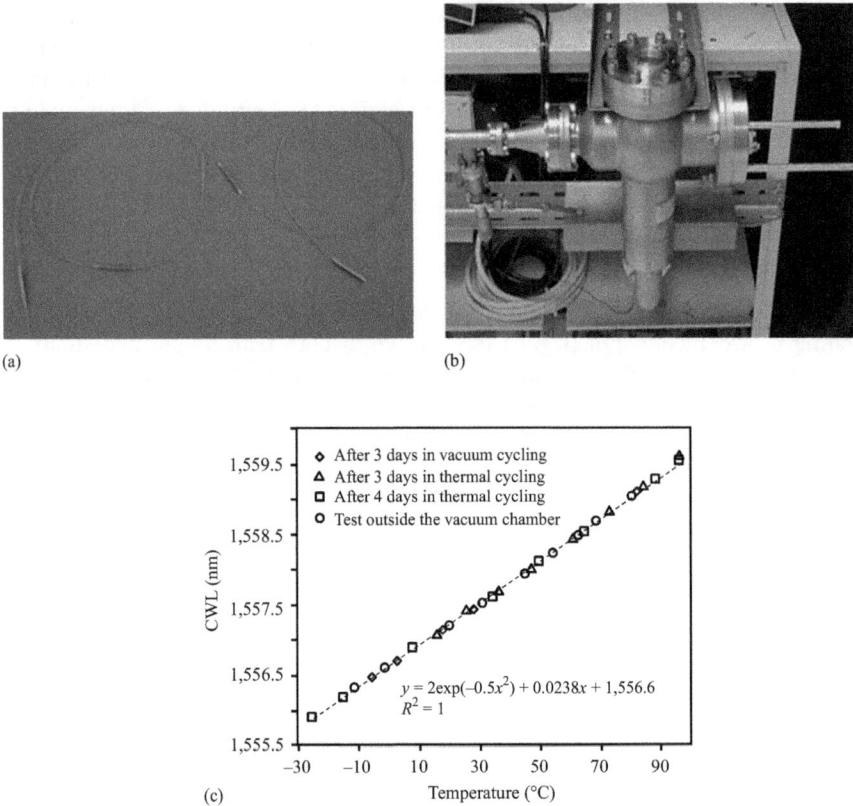

Figure 7.15   *(a) Photograph of a packaged flight FBG T-sensor line with*
protective cabling and four sensors, (b) vacuum chamber (<10 Torr
using Turbo pump, silica fibre vacuum feed through) and (c) vacuum
thermal cycling tests for FBG temperature sensor: test results for
one of the FSD FBG T-sensors (photos courtesy of MPB
Communications, Inc.)

The high temperature-sensor 3 (*T*-3) was mounted on a small shaker. It then underwent 6 h of continuous sinusoidal vibration cycling. Each cycle consisted of a sinusoidal sweep from 2 to 2,000 Hz in 15 min. This was then repeated 25 times.

Following the vibration testing, the T-3 FBG RT wavelength was basically unchanged from the wavelength prior to the start of the sinusoidal vibration testing. The sensor was then cycled from RT to $+400°C$ several times in air using a small ceramic oven with no ill effects. Some of the calibration and test results are shown in Figure 7.17(c). This sensor is still fully operational to high temperatures despite considerable handling, vibration and thermal cycling.

Representative high *T*-sensor, FBG *T*-sensor and *P/T* sensor cables assemblies were shipped to SSTL and integrated with the PROBA-2 propulsion subsystem.

(a)

(b)

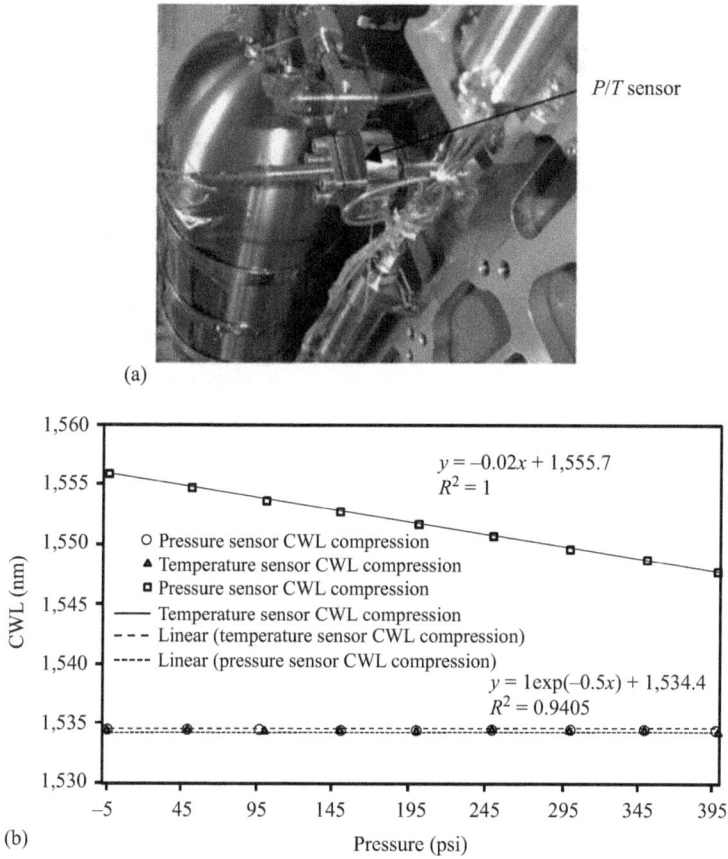

Figure 7.16 *(a) P/T sensor attached to pressurisation system for sensor testing using N_2 and (b) measured characteristics for the P/T-sensor, P-senor and T-sensor FBG elements at different applied pressures: pressure calibration of the P/T sensor (125.5 μm flight) after calibration and potting*

Figure 7.17(a) shows the proto high *T*-sensor mounted on the SSTL thruster. Also, for comparison, is the standard thermocouple with SS coating. The representative sensors, integrated with the SSTL propulsion system, successfully underwent the random vibration testing to the PROBA-2 levels.

7.5.3 System ground qualification

The initial system environmental testing was employed to improve the PCB and fibre optic component mounting. The initial testing included

- RS-422 data interface communication,
- full functionality test,

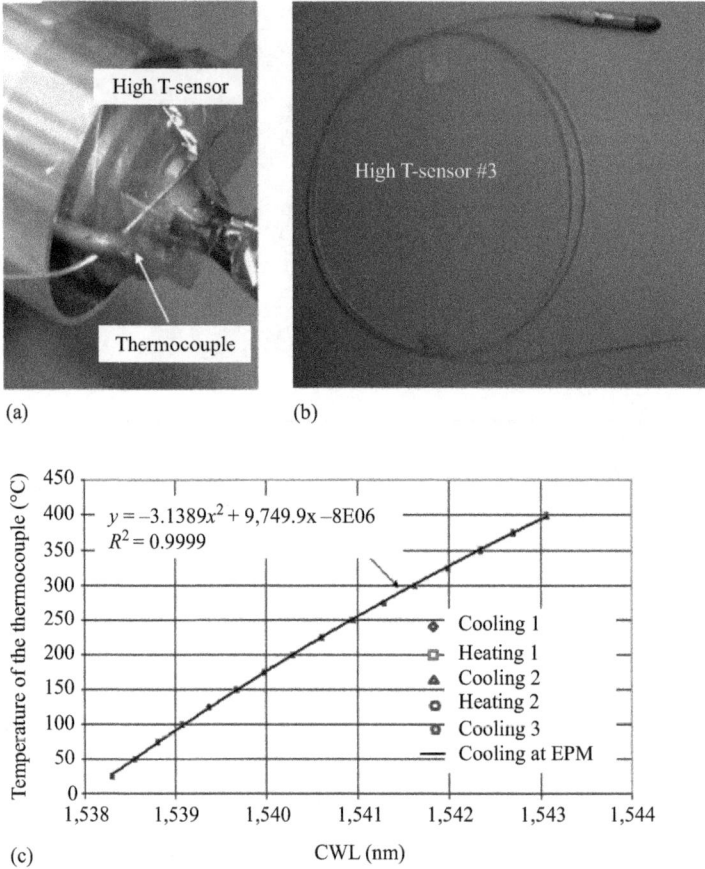

Figure 7.17   (a) Preliminary integration of high T-sensor with the SSTL thruster,
(b) photograph of high T-sensor cable with FC/APC optical
connector and (c) high T-sensor calibration: cooling 1, heating 1 and
cooling 2 were performed during the same day, while the heating 2
and cooling 3 were performed 3 days later. A last cooling test was
performed at Ecole Polytechnique de Montreal (EPM) for
comparison purpose (courtesy of MPB Communications, Inc.)

• fibre laser operation between −20°C and +40°C using an environmental chamber,
• low-level 5–2,000 Hz sine test for resonances,
• random sinusoidal vibrations at 16.3 g and
• operation in thermal vacuum between 0°C and 40°C.

The operation of the tuneable fibre laser was observed while subjecting the fibre laser to moderate level sinusoidal vibrations between 2 and 2,000 Hz (Figure 7.18). It was found that the fibre laser tuning was still operational even with the moderate sinusoidal vibrations applied to the tuneable filter during the operation. Table 7.4 shows the random vibration testing of the FSD representative unit in two separate sections, namely the optical tray and FSD unit with mounted PCBs on aluminium stiffener.

The same spectrum, as shown in Figure 7.19, was applied to all the tests for the optical tray in the vertical direction (Z), and for the FSD electronics mounted within the enclosure in all the directions X, Y and Z. The test duration in each axis was 1.5 min. Before and after each random vibration test, a low amplitude (1 g)

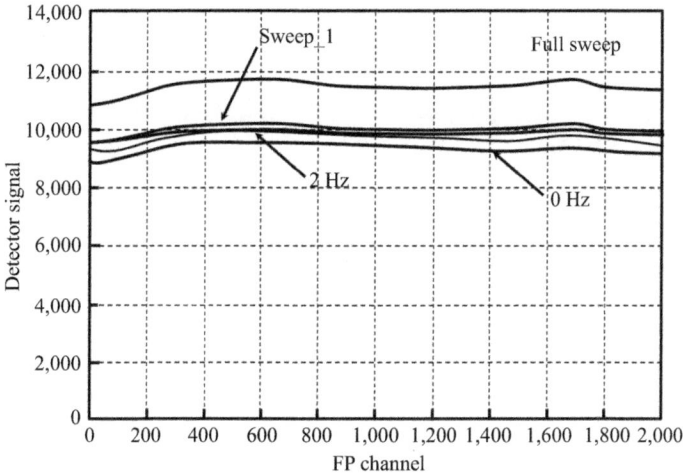

Figure 7.18   Measured signal output of tuneable fibre laser in the 1,520–1,560 nm spectral range during moderate sinusoidal vibration at several g. The signal was smoothed with Sevitzky-Golay method over 20 point of window

Table 7.4   Random vibration testing of the FSD representative unit in two separate sections: optical tray, and FSD unit with mounted PCBs on Al stiffener

Testing	Part	Measurements	Test
Mechanical	Optical tray with one attached FBG sensor line	Performance functional test before and after each vibration test	• Low level sinusoidal vibration scan between 20 and 2,000 Hz • Random vibration in the three axis
Mechanical	FSD enclosure with CPU and EO/DAQ PCBs on stiffener	Performance functional test before and after each vibration test	• Low level sinusoidal vibration scan between 20 and 2,000 Hz • Random vibration in the three axis

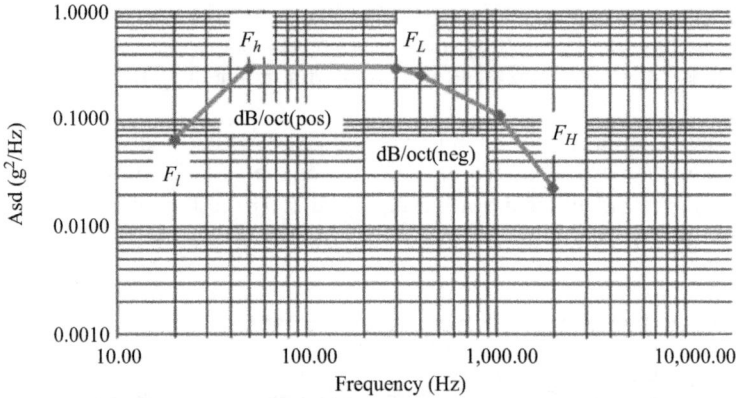

		Slope				Acceleration
Freq (Hz)	Asd (g²/Hz)	dB	oct	dB/oct	Area	Grms
20	0.0629	*	*	*	*	*
50	0.3	6.79	1.32	5.13	5.08	2.25
300	0.3	0	2.58	0	80.08	8.95
400	0.2571	−0.67	0.42	−1.61	107.78	10.38
1,050	0.1086	−3.74	1.39	−2.69	212.33	14.57
2,000	0.0229	−6.77	0.93	−7.28	260.48	16.14

Figure 7.19   Random vibration spectrum for FSD enclosure applied in all directions, with total of 16.14 g (Asd: acceleration spectral density) F_l and F_h: lower and higher frequencies at the positive Asd slope, and F_L and F_H are the lower and higher frequencies at the negative Asd slope, respectively

sinusoidal vibration test, between 20 and 2,000 Hz at 4 octet/min, was affected (pre-sine and post-sine). The pre-sine and post-sine provide the spectrum response of the tested part and indicate its natural frequency and the vibration amplification The comparison of the pre-sine and post-sine amplifications at different frequencies informs on the changes in the structure (even not seen) through the change of the unit natural frequency and its amplitude.

Since the largest unknown was the effect of the high-level random vibration on the fibre optic components, a separate test was performed using just the optical tray and mounted fibre optic components. This comprised the base section of the interrogation unit. Special potting was employed for the fibre optic component mounting to help dampen the effects of vibrations as well as to minimise thermal stress on the integral optical fibres. The FSD optical tray was tested only in the Z direction. The main risk is that the fibre or optical component would move in the vertical axis relative to the base of the optical tray. Figure 7.20 shows a photograph of the FSD enclosure with the PCB attached to the adapter (x-axis vibrations) and mounted on the CSA shaker.

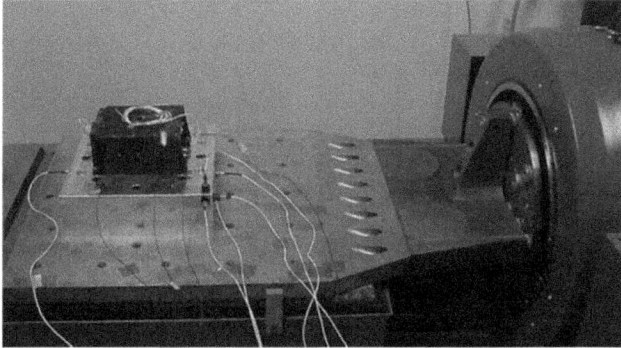

Figure 7.20   FSD enclosure with the PCB attached to the adapter (x-axis vibrations) and mounted on the CSA shaker

For the testing of the FSD interrogation unit, four accelerometers were attached to the FSD enclosure:

- An accelerometer in the X direction on the lateral side of the FSD enclosure.
- An accelerometer in the Y direction on the lateral side of the FSD enclosure.
- An accelerometer in the Z direction, inside the FSD enclosure on the upper circuit board, on the top of the DC–DC converter, then the enclosure was closed.
- An accelerometer in the Z direction on the top of the FSD enclosure.

The optical tray and mounted components exhibited no undue resonances between 20 and about 2,000 Hz. The fibre laser system was fully functional after the high-level random sine vibration test. The only slight resonance of note was near 1,800 Hz for the electronics PCB assembly, mainly at the DC–DC converter that was unstrapped for the initial random vibration testing. Functionality testing was performed after each environmental stress. The FSD unit was fully functional after the completion of the random vibration testing.

Following the mechanical vibration testing, the representative FSD unit was mounted in the thermal vacuum chamber at CSA. Its operation in vacuum as tested between 0°C and 40°C. At the 40°C ambient temperature, the FSD electronic components were still well within their nominal temperature ranges. Moreover, the tunable fibre laser was fully functional at the temperature extremes.

7.6   Fibre optic sensor demonstrator flight and validation

The FSD flight unit testing was accomplished at CSA's David Florida laboratories in Ottawa, Ontario. The testing included

- low-level 5–2,000 Hz sine test for resonances,
- random sinusoidal vibrations at 16.3 g,
- system operation in thermal vacuum between −40°C and +60°C and
- EMC/EMI testing.

Functionality testing was performed after each environmental stress. The FSD unit was fully functional after the completion of the random vibration testing.

Following the mechanical vibration testing, the FSD flight unit was mounted in the thermal vacuum chamber. The FSD operation in vacuum was tested at $-40°C$, $-20°C$, $0°C$, $20°C$, $40°C$ and $60°C$.

7.6.1   Interrogation with PROBA-2

The FBG temperature sensors were routed loosely from the central interrogation system, respecting a minimum bend radius of 20 mm, to minimise any stress on the cabling from thermal expansion/contraction. The temperature sensors were attached at the desired points using a combination of thermal epoxy and aluminium tape, as shown in Figure 7.21. The fibre cabling was secured every 5–6 cm using aluminium tape.

Two redundant lines, TS-003 and TS-004, are employed to measure the temperature at critical points along the SSTL xenon pipeline, as shown in Figure 7.22.

The FSD employs a redundant temperature-compensated narrowband reference FBG in series with the external FBG sensors for each of the sensor lines to provide an absolute wavelength reference. The temperature dependence of the reference FBG's is about 0.001 nm/°C. The temperature of the reference FBG aluminium holder is monitored using two AD590 thermistors. This was used to correct for the slight wavelength shift of the reference FBG's with temperature.

The sensor signals typically exceed 1 V at the peak value relative to a noise level below 0.1 mV to provide a high-measurement SNR ($>10^4$). This can accommodate a relatively large signal intensity decrease exceeding 10 dB at the sensor system end-of-life. This large margin should facilitate high performance for extended duration space missions.

The high T-sensor tip was inserted in a small aluminium adapter (Figure 7.23) that had been previously affixed by SSTL to its thruster. The sensor was secured using Durabond 950 high temperature aluminium-based thermal epoxy.

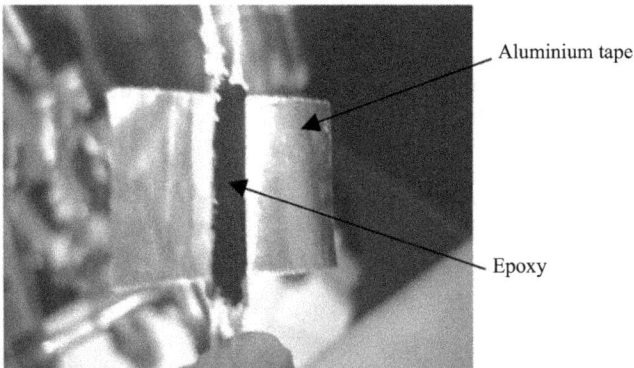

Figure 7.21   FBG sensor mounting using combination of thermal epoxy and aluminium tape

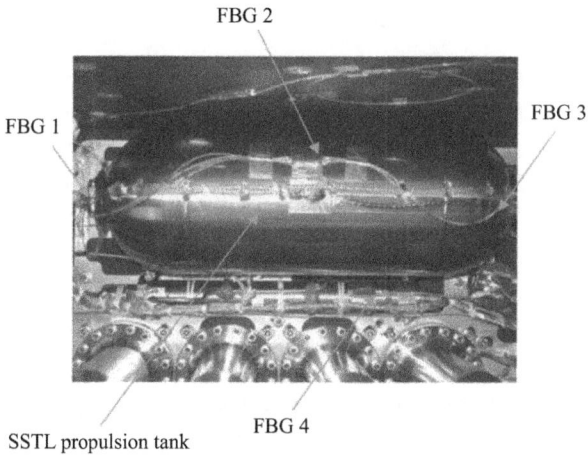

Figure 7.22 *Photograph of FBG sensors integrated with SSTL propulsion subsystem*

Figure 7.23 *FBG high T-sensor integrated with SSTL thrusters using aluminium adapter piece*

7.7 Experimental high-temperature fibre Bragg grating regeneration and sensor packaging

The first time FBGs were brought and kept at temperatures above 700°C, their signal intensities underwent considerable decay and then completely disappeared, and then went through a regeneration process (Figure 7.24). The regenerated FBG CWL were 1.5–2 nm shifted, and their intensities were lower than the original ones. The optimal temperature for regeneration was found to be around 900°C–930°C, and the regeneration process took a few hours. The regenerated FBG response to temperature was calibrated with thermocouples using standard cylindrical ceramic oven. A preliminary test confirmed the stability of the FBG intensity and CWL of the sensors at 1,000°C for 24 h.

The regenerated FBG signal stayed up to about 1,350°C. However the fibre became fragile after the coating gold melted at about 1,050°C. The regeneration phenomenon is still not well understood, and various groups are currently studying it [79].

Fibre sensors based on FBG are stable for temperatures up to 1,000°C using gold-coated fibres with special packaging optimising between protective capability for the passive fibre and fast thermal conductivity on the sensor locations.

To select the optimal packaging design of the fibre sensors, a series of fibres were prepared and tested in different small tubes of less than 1 mm output diameter. The optimal option was found to install only one fibre within small thin SS tube of 0.35/0.45 mm input/output diameters. This design permitted to follow fast transient with a thin wall and at the same time offering a wall strong enough to prevent the tube buckling. An example of a packaged gold-coated fibre sensor is shown in Figure 7.25.

7.7.1   Validation test at re-entry environment plasmas laboratories

Figure 7.26 shows a photograph of the plasma hitting the centre of the probe with the FBG sensor and the photograph obtained by the IR camera. Representative fibre sensor package was attached to two thermal protection system materials, one

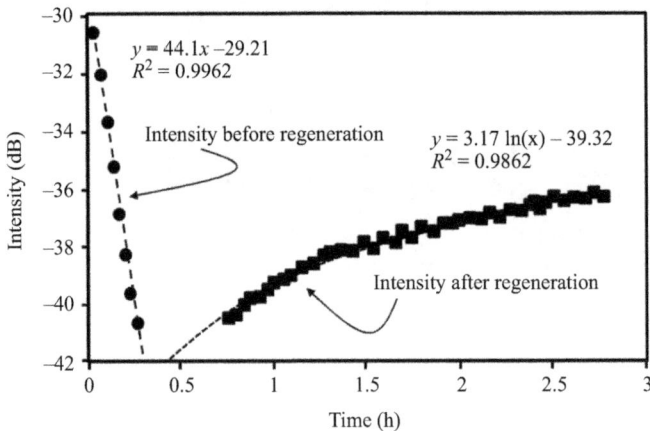

Figure 7.24   Evolution of intensity of the FBG sensor set at 900°C during regeneration. The peak disappears for about half an hour, before starting to be regenerated

Figure 7.25   Example of a packaged gold-coated FBG sensor

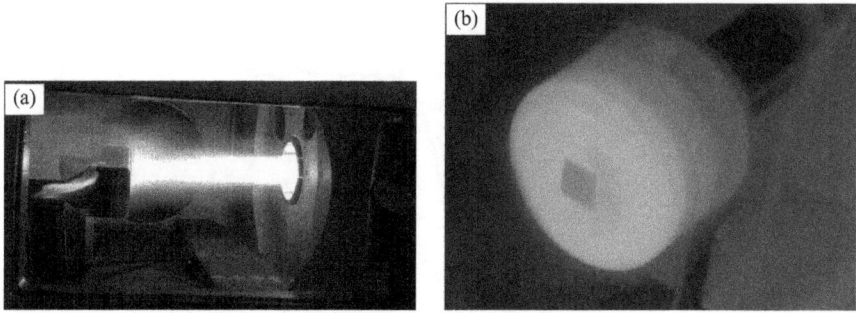

Figure 7.26 *Photograph of (a) the plasma hitting the centre of the probe with the FBG sensor. The second adjacent is the heat flux probe and (b) photograph provided by the IR camera*

metallic (Hastalloy^{TM} 230, flexible, heavy) and one ceramic (C/SiC higher temperature, smaller mass, brittle). These prototypes were validated at Von Karman Institute (Belgium), plasmatron in a harsh environment similar to atmospheric re-entry plasma. The plasma monitoring included:

- one thermocouple on the sample under test, close to the FBG packaging;
- an IR camera;
- a pyrometer and
- a heat flux probes were measuring the plasma temperature and the heat flux, at the stagnation point.

Three fibre interrogators were used to register the FBG temperature dynamics as follows:

- a Burleigh wavemeter, at 0.2 Hz,
- an interrogator similar to that flying on PROBA-2 and
- A commercial FOSs and sensing systems module.

The FBG sensors were heated and successfully sensed temperatures above 1,100°C. The gold-coating melted at around 1,050°C, and the fibres without their protective coating survived for 5 min before breaking under the vibrations. Figure 7.27 compares the sensing temperature result by four instruments monitoring the temperature. The thermocouple and FBG sensor (Burleigh measurements) provided very similar values, whereas the recorded values by the pyrometer and the IR camera temperatures were slightly higher as they were located in different areas.

7.7.2 *Validation Test in a wind tunnel – Deutsches Zentrum für Luft- und Raumfahrt, Cologne*

A second validation of the fibre sensors was performed in a plasma wind tunnel at Deutsches Zentrum für Luft- und Raumfahrt (DLR) in Cologne, Germany, simulating the re-entry at 8 Mach number (3.8 km/s) and 1,000°C (Figure 7.28). The device tested was a (C/SiC) tile, 150×250×3 mm thick, similar to those installed

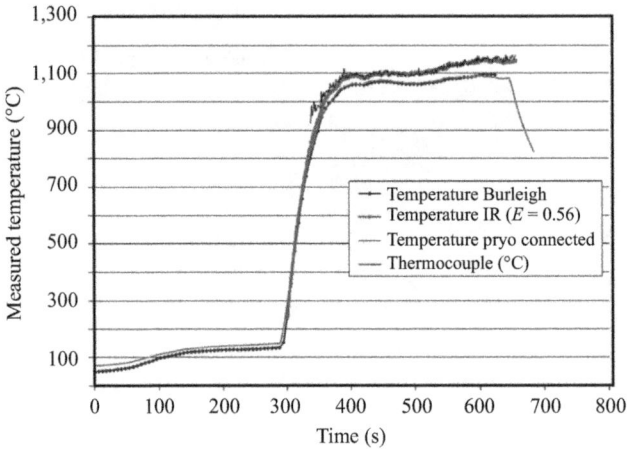

*Figure 7.27   Comparison of the plasma temperature as measured by the FBG,
thermocouple, pyrometer and IR camera*

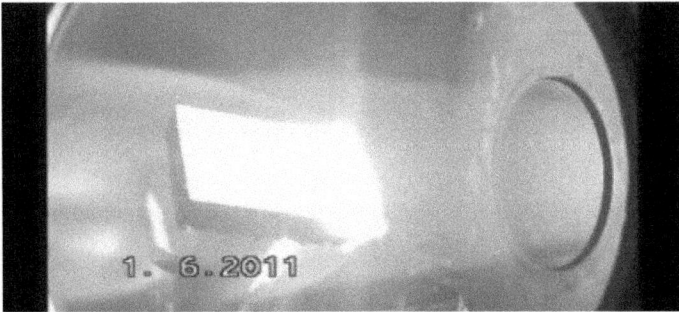

*Figure 7.28   Picture of the C/SiC with the FBG and thermocouples during the
wind tunnel test at DLR agency*

on SHEFEX series re-entry vehicles built by DLR. The tile embedded three fibre
lines, each of them containing three sensors, in addition to a dozen of thermocouples that permitted to show the similarity between the transients and plateaus
temperatures seen by the FBG (at 100 Hz frequency acquisition rate) and those seen
by the thermocouples (10 Hz frequency acquisition rate).

Figure 7.29 presents the transient temperatures and plateau as seen by three
thermocouples around FBG sensors. The thermocouples are located in slightly
deeper grooves (0.3 mm) compared to the fibre sensors, which makes them sense
the transients slightly earlier. The response time of the FBG is faster than 0.5 µs, as
demonstrated in hypervelocity impacts study which compares the fibre sensor
response with the standard SG sensors.

Some of the FBG were at 1,020°C and 1,030°C and stayed completely intact
during all wind tunnel tests and temperature ramp down.

7.7.3 ROTEX-T re-entry mission

Containing our FBG sensors, ROcket Technology EXperiment-Transition (ROTEX-T) is an experimental rocket which was launched in August 2016. The extension passive fibre between the sensors and the interrogator was found to break inside the motor adapter during the rollout to the launch stage. It was not possible to have access to the launch stage for security reason.

During the complete flight, although it could not record any temperature, the interrogator itself worked normally and the telemetry data was received [79].

The ROTEX-T payload part impacted in the target area with a velocity of about 95 m/s. Figure 7.30 shows the interrogator and the USB after being mounted

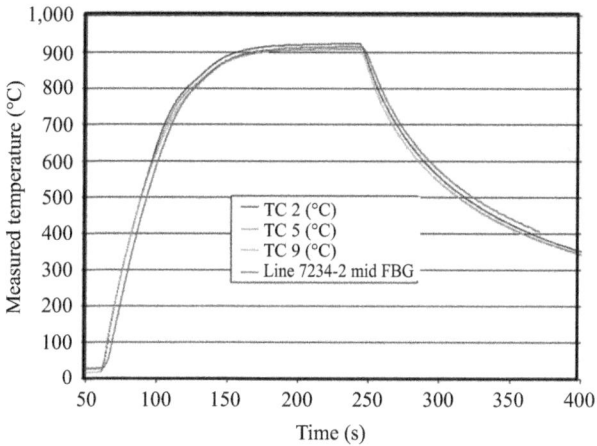

Figure 7.29   Measured temperatures by one FBG sensor and a thermocouple (TC) during the wind tunnel test at DLR agency

Figure 7.30   Interrogator enclosure and the USB box (a) after attachment to the ROTEX-T internal payload and (b) as recuperated after the re-entry and the on-ground impact

Figure 7.31   Details of the interrogator front with the connectors. Three of them having FBG are found to be still attached

on the internal ROTEX-T set-up, and as recuperated after the re-entry experiment and on ground impact. The housing of interrogator and memory stick box were undamaged (some of the mounting feet were slightly deformed).

All our four FBG connectors were still attached Figure 7.31; one was bent with no fibre, the other three had about 1.5 m fibre extension (probably this is where the fibres broke during the rollout).

The internal memory card was removed and was completely functional as tested at DLR. The card was still working normally and the flight data files were successfully downloaded. The reading was correct although no temperatures could be recorded.

7.7.4   Measuring experimentally the stability of fibre Bragg gratings under high gamma radiation

Finally, a high radiation test addressing the stability of FBG in equivalent harsh radiation space environment was performed on three kinds of packaging:

- bare FBG (only 5 μm polyimide coating),
- medium thickness tube SS (0.83 mm thick) and
- larger thickness SS tube (1.37 mm thick).

The FBGs were submitted to high radiation level (25 Mrad at 420 krad/h), 250 times larger than the 100 krad commonly requested for a 10 years satellite lifetime. The effect on the FBG was a small shift (barely 200 pm, equivalent to the shift in the CWL).

7.8   Conclusions

Spacecraft require extensive *in situ* monitoring of their status and thermal performance, both during ground validation and the subsequent mission in the space environment. The fibre sensors demonstrate their advantages in space applications for temperature and strain measurements. A full FOS network in the space environment on a satellite was demonstrated in PROBA-2. They show high resistance against gamma radiation. Tests during atmospheric re-entry demonstrated the fibre suitability to follow this fast dynamic transition. Smart satellite panels with integrated optical fibres for sensing and data communication are being built. The *ad hoc* assembly of various electronic sensors and processing electronics at a substantial mass and performance penalty due to the EMI sensitivity and resultant shielding requirements. Replacing the classical electrical system with fibre optic systems save a factor of 6 on the mass of the electronics (power source, data acquisition and processing), a factor of 6 on the mass of sensors (fibres sensors *versus* thermocouples and thermistors) and a factor of 1.5 for the time needed to integrate the sensors within the structure.

The SL interrogator system developed for a future FBG-sensing system on board a launcher or another spacecraft. Due to its compact all-in-fibre set-up based on laser diode, no free space or moving parts are necessary for sensor interrogation. The power supply, control and signal measurement electronic are usually based on commercially available components for the breadboard development of the SL. Since the requirements for radiation hardness of components are low due to the short lifetime of launchers, the development is focused on system compliance with the required measurement performance:

- The primary application for a future SL would be the temperature measurement acquiring housekeeping data. Therefore, the SL is best suited because of its broad optical output spectrum of more than 40 nm that allows to connect a high number of sensors to be measured sequentially. Since the measurement rate for temperature sensing is usually limited to less than 10 Hz, up to several hundred FBG sensors can be measured by future versions of the SL interrogator. Although the SL interrogator has demonstrated reliable measurement results, further development is necessary to come to a system suitable for spacecraft. The impact of space environment to the laser diode has not been investigated until now.
- FSD was the first demonstration of a full FOS network in the space environment on a satellite. The initial ground random vibration and thermal vacuum tests results indicate good performance for the FSD selected fibre optic components, innovative FBG sensors and fibre optic cabling. The FSD technology demonstrator has been designed to showcase the overall advantages of FOS for space systems. The initial ground random vibration and thermal-vacuum test results indicate good performance for the FSD-selected fibre optic components, innovative FBG sensors and fibre optic cabling.

References

[1]   I. McKenzie and N. Karafolas, *Proceedings of SPIE*, 2005, **5855**, 262.
[2]   E. Friebele, C.G. Askins, A.B. Bosse, *et al.*, *Smart Materials and Structures*, 1999, **8**, 813.

[3]   R. Kruzelecky, J. Zou, E. Haddad, *et al.*, *Proceedings of the 7th International Conference on Space Optics*, Toulouse, France, 14–17 October 2008.

[4]   R. Kruzelecky, J. Zou, N. Mohammed, *et al.*, *Proceedings of the 7th International Conference on Space Optics*, Toulouse, France, 14–17 October 2008.

[5]   A. Othonos and K. Kalli in *Fibre Bragg Gratings: Fundamentals and Applications in Telecommunications and Sensing*, Artech House, Norwood, MA, USA, 1999.

[6]   F.T. Yu and Y. Shizhuo in *Fibre Optic Sensors*, Marcel Dekker Inc., New York, NY, USA, 2002.

[7]   K. Hill and G. Meltz, *Journal of Lightwave Technology*, 1997, **15**, 8, 1263.

[8]   A. Kersey, M.A. Davis, H.J. Patrick, *et al.*, *IEEE Journal of Lightwave Technology*, 1997, **15**, 8, 1442.

[9]   OZ Optics in *ASE Broadband Light Source*, OZ Optics, March 2010. *https:// www.ozoptics.com/ALLNEW_PDF/DTS0106.pdf.*

[10]   J.O. Wesstrom, S. Hammerfeldt, J. Buus, R. Siljan, R. Laroy and H. de Vries, *Proceedings of the 18th IEEE International Semiconductor Laser Conference*, Kongresshaus Garmisch-Partenkirchen, Garmisch, Germany, 29 September–3 October 2002, 99.

[11]   K. Tsuzuki, Y. Shibata, N. Kikuchi, *et al.*, *IEEE Selected Topics in Quantum Electronics*, 2009, **15**, 3, 521.

[12]   M. Bonesi, M.P. Minneman, J. Ensher, *et al.*, *Optics Express*, 2014, **22**, 3, 2632.

[13]   C. Hawthorn, K. Weber and R. Scholten, *Review of Scientific Instruments*, 2001, **72**, 12, 4477.

[14]   M.S. Müller, L. Hoffmann, T. Bodendorfer, *et al.*, *IEEE Transactions on Instrumentation and Measurement*, 2010, **59**, 3, 696.

[15]   P. McNamara, S. Vitale and K. Danzmann, *Classical and Quantum Gravity*, 2008, **25**, 11.

[16]   European Space Agency in *Technology Readiness Levels Handbook for Space Applications*, European Space Agency, September 2008. *https://artes. esa.int/sites/default/files/TRL_Handbook.pdf.*

[17]   J. Mankins in *Technology Readiness Levels*, Advanced Concepts Office, Office of Space Access and Technology NASA, April 1995. *http://www. artemisinnovation.com/images/TRL_White_Paper_2004-Edited.pdf.*

[18]   W. Ley, W. Wittmann and W. Hallmann in *Handbuch der Raumfahrttechnik*, Hanser, Munich, Germany, 2008.

[19]   R.D. Williams, B.W. Johnson and T.E. Roberts, *IEEE Micro*, 1988, **8**, 4, 18.

[20]   F.L. Doudkin and M.P. Gough, *Radio Science*, 1999, **34**, 5, 1299.

[21]   W. Paul, *Review of Modern Physics*, 1990, **62**, 531.

[22]   W. Kriedte, *Proceedings of the Spacecraft Structures, Materials & Mechanical Testing Conference*, Noordwijk, The Netherlands, 27–29 March 1996, 321.

[23]   D. Harland and R. Lorenz in *Space Systems Failures – Disasters and Rescues of Satellites, Rockets and Space Probes*, Springer, Berlin, Germany, 2005.

[24] C. Dyer and G. Hopkinson in *Space Radiation Effects for Future Technologies and Missions*, QinetiQ Space Department, Chertsey, Surrey, UK, 2001.

[25] J. Barth, C. Dyer and E. Stassinopoulos, *IEEE Transactions on Nuclear Science*, 2003, **50**, 3, 466.

[26] A. Holmes-Siedle and L. Adams in *Handbook of Radiation Effects*, Oxford Science Publications, New York, NY, USA, 1993.

[27] E. Daly, P. Nieminen, A. Mohammadzadeh, *et al.*, *Proceedings of the 7th on Radiation and its Effects on Components and Systems (RADECS)*, Noordwijk, The Netherlands, 2003, 175.

[28] J. Srour and J. McGarrity, *Proceedings of the IEEE*, 1988, **76**, 11, 1443.

[29] T. Oldham and F. McLean, *IEEE Transactions on Nuclear Science*, 2003, **50**, 3, 483.

[30] S.T. Johns, M.J. Hayduk, R.J. Bussjager, *et al.*, *Electronics Letters*, 2003, **39**, 18, 1310.

[31] R. Pease, *IEEE Transactions on Nuclear Science*, 2003, **50**, 3, 539.

[32] A. Johnston, B. Rax and C. Lee, *IEEE Transactions on Nuclear Science*, 1995, **42**, 6, 1650.

[33] A. Johnston, B. Rax and C. Lee, *IEEE Transactions on Nuclear Science*, 1996, **43**, 6, 3049.

[34] J. Titus, D. Emily, J. Krieg, T. Turflinger, R.L. Pease and A. Campbell, *IEEE Transactions on Nuclear Science*, 1999, **46**, 6, 1608.

[35] D.M. Fleetwood, S.L. Kosier, R.N. Nowlin, *et al.*, *IEEE Transactions on Nuclear Science*, 1994, **41**, 6, 1871.

[36] D. Schmitt-Landsiedel in *Elektronische Bauelemente*. Lehrstuhl für Technische Elektronik, 2003.

[37] E. Taylor, K. Hulick, J.M. Battiato, A.D. Sanchez, J.E. Winter and A. Pinch, *Proceedings of the Photonics for Space and Radiation Environments*, EUROPTO Series, Florence, Italy, 20–21 September 1999, 72.

[38] U. Keller, *Nature*, 2003, **424**, 831.

[39] A. Ellis in *Teaching General Chemistry: A Materials Science Companion*, American Chemical Society, Washington, DC, USA, 1993.

[40] K. Nassau, *American Mineralogist*, 1978, **63**, 219.

[41] C. Wright in *Spectroscopic Characterization of Fluorite: Relationships Between Trace Element Zoning, Defects and Color*, Miami University, Miami, FL, USA, 2002.

[42] M.N. Ott in *Radiation Effects Data on Commercially Available Optical Fibre: Database Summary*, Sigma Research and Engineering, NASA Goddard Space Flight Center, Greenbelt, MD, USA, 2002. *https://photonics.gsfc. nasa.gov/tva/meldoc/W-5datawkshp2002.pdf*.

[43] M.N. Ott, *Photonics for Space Environments VI, Proceedings of SPIE Vol. 3440*, 1998, 37–46.

[44] M.N. Ott in *Space Flight Applications of Optical Fibre; 30 Years of Space Flight Success*, NASA Goddard Space Flight Center, Greenbelt, MD, USA. *https://pdfs.semanticscholar.org/c533/ee4324b04a99f5ccd5b7aec5840ea4520 a32.pdf*.

[45]   M.N. Ott in *Radiation Effects Expected for Fibre Laser/Amplifier Rare Earth Doped Optical Fibre*, NASA Goddard Space Flight Center, Greenbelt, MD, USA, March 2004. *https://photonics.gsfc.nasa.gov/tva/meldoc/fiberla serradiationeffects.pdf.*

[46]   M. Caussanel, Ph. Signoret, O. Gilard, M. Sotom, A. Touboul and J. Gasiot, *Proceedings of the 7th on Radiation and its Effects on Components and Systems (RADECS)*, Noordwijk, The Netherlands, 2003, 553.

[47]   S. Girard, Y. Ouerdane, B. Tortech, *et al.*, *IEEE Transactions on Nuclear Science*, 2009, **56**, 6, 3293.

[48]   P. Borgermans, B. Brichard, F. Berghmans, *et al.*, *Proceedings of the SPIE on Fibre Optic Sensor Technology II*, 2001, **4204**, 151.

[49]   B. Brichard, P. Borgermans, F. Berghmans, *et al.*, *Proceedings of the SPIE on Photonics for Space and Radiation Environments*, 1999, **3872**, 36.

[50]   B. Brichard, A.F. Fernandez, H. Ooms, P. Borgermans and F. Berghmans, *Proceedings of SPIE on the 2nd European Workshop on Optical Fibre Sensors*, 2004, **5502**, 184.

[51]   B. Brichard, A.F. Fernandez, O. Alberto, B. Hans, F. Berghmans, *Proceedings of the 7th on Radiation and its Effects on Components and Systems (RADECS)*, Noordwijk, The Netherlands, 2003.

[52]   K. Predehl in *Frequenzkämme für Weltraumanwendungen*, Ludwig Maximilian University of Munich, Munich, Germany, 2006. [Diploma Thesis] [in German].

[53]   M. Lezius, K. Predehl, W. Stower, *et al.*, *IEEE Transactions on Nuclear Science*, 2012, **59**, 2, 425.

[54]   H. Henschel, O. Kohn, H.U. Schmidt, J. Kirchof and S. Unger, *IEEE Transactions on Nuclear Science*, 1998, **45**, 3, 1552.

[55]   E.W. Taylor, S.J. McKinney, A.D. Sanchez, *et al.*, *Proceedings of SPIE on Conference of Photonics for Space Environments VI*, 1998, **3440**, 16.

[56]   B. Fox, Z.V. Schneider, K. Simmons-Potter, *et al.*, *Proceedings of SPIE on Conference of Fibre Lasers IV: Technology, Systems, and Applications*, 2007, **6453**, 645328.

[57]   H. Henschel, O. Köhn and U. Weinand, *IEEE Transactions on Nuclear Science*, 2002, **49**, 3, 1401.

[58]   H. Henschel, J. Kuhnhenn and U. Weinand, *Proceedings of the Optical Fibre Communication Conference*, Anaheim, CA, USA, 6–11 March 2005.

[59]   K.V. Zotov, M.E. Likhachev, A.L. Tomashuk, *et al.*, *Proceedings of the 9th European Conference on Radiation and its Effects on Components and Systems (RADECS)*, 2007, doi:10.1109/RADECS.2007.5205517.

[60]   K.V. Zotov, M.E. Likhachev, A.L. Tomashuk, *et al.*, *IEEE Photonics Technology Letters*, 2008, **20**, 17, 1476.

[61]   F. Berghmansa, A.F. Fernandeza, B. Bricharda, *et al.*, *Proceedings of SPIE International Symposium on Industrial and Environmental Monitors and Biosensors Harsh Environment Sensors*, 1998, **3538**, 28.

[62]   A.I. Gusarov, F. Berghmans, O. Deparis, *et al.*, *IEEE Photonics Technology Letters*, 1999, **11**, 9, 1159.

[63] E.W. Taylor, *Proceedings of the 1999 IEEE Aerospace Conference*, 1999, **3**, 307.

[64] A. Gusarov, D. Kinet, C. Caucheteur, M. Wuilpart and P. Megret, *IEEE Transactions on Nuclear Science*, 2010, **57**, 6, 3775.

[65] A. Gusarov, B. Brichard and D. Nikogosyan, *IEEE Transactions on Nuclear Science*, 2010, **57**, 4, 2024.

[66] A. Gusarov, C. Chojetzki, I. Mckenzie, H. Thienpont and F. Berghmans, *IEEE Photonics Technology Letters*, 2008, **20**, 21, 1802.

[67] A. Fernandez, B. Brichard and E. Berghmans, *Proceedings of the 7th on Radiation and its Effects on Components and Systems (RADECS)*, Noordwijk, The Netherlands, 2003.

[68] A. Gusarov and S.K. Hoeffgen, *IEEE Transactions on Nuclear Science*, 2013, **60**, 3, 2037.

[69] A. Othonos, *Review of Scientific Instruments*, 1997, **68**, 4309.

[70] European Space Agency in *The Space Environment Information System*, SPENVIS, European Space Agency, 2009. *http://www.spenvis.oma.be/*.

[71] G. Wrenn, D. Rodgers and K. Ryden, *Annales Geophysicae*, 2002, **20**, 953.

[72] J. Scialdone, P. Isaac, C. Clatterbuck and R. Hunkeler in *Material Total Mass Loss in Vacuum Obtained from Various Outgassing Systems*, NASA Goddard Space Flight Center, Greenbelt, MD, USA, July 2000.

[73] E. Perez in *Ariane 5-User's Manual*, 5th Edition, Arianespace, Cedex, France, July 2008.

[74] Y.J. Rao, *Optics and Lasers in Engineering*, 1999, **31**, 297.

[75] R. Brandt, *Proceedings of the Spacecraft Propulsion International Conferences*, Toulouse, France, 8–10 November 1994.

[76] J. Rao, *Measurement Science and Technology*, 1997, **8**, 355.

[77] A. Rogers, *Measurement Science and Technology*, 1999, **10**, R75.

[78] M.N. Ott and P. Friedberg in *Technology Validation of Optical Fibre Cables for Space Flight Environments*, Sigma Research and Engineering, NASA Goddard Space Flight Center, Greenbelt, MD, USA, 2000.

[79] E. Haddad, R.V. Kruzelecky, K. Tagziria, *et al.*, *Proceedings of the International Conference on Space Optics (ICSO)*, Paper No. 200, Biarritz, France, 18–21 October 2016.

Chapter 8

Fibre Bragg gratings/microelectromechanical system-integrated optical devices

In this chapter, we describe the integration of fibre-optic signal distribution with fibre Bragg gratings (FBG) and microelectromechanical systems (MEMS) actuators to facilitate various miniature tunable fibre-optic devices for applications ranging from fibre lasers, to variable time delay lines for optical signal processing, to the sensing of various physical parameters such as temperature, pressure or acceleration including our own most recent results.

8.1 Introduction

Fibre-optic sensors (FOS) employ a signal link *via* an optical fibre, allowing the subsequent electronic processing to be located remotely from critical areas of a facility or spacecraft. Signals on a fibre-optic line are bidirectional, allowing a single fibre to carry both the source signal to the optical sensor and the reflected return signal back to the interrogation system. Due to the low signal loss, <1 dB/km, relatively long signal links are feasible with good signal integrity.

FOS are typically insensitive to electromagnetic interferences (EMI), allowing relatively simple fibre-optic signal routing and allowing a low-mass signal harness due to the reduced shielding requirements, mainly for mechanical protection.

The all-optical signal distribution can assist to avoid safety issues, such as sparking, in critical application areas such as spacecraft propulsion subsystems, chemical production and grain storage.

The benefits of a FOS network include EMI insensitivity, remote positioning of the interrogation system at some distance from the sensors, flexible signal routing using lightweight fibre-optic cables with armoured microtubing for high mechanical strength, high sensor capacity per fibre-optic line using wavelength-division multiplexing (WDM), and high measurement resolution of the sensor response. Figure 8.1 summarises different types of FBG, including linear and chirped Bragg gratings, Fabry–Pérot (FP) cavity grating and long-period grating.

The fibre optical solution can provide substantial benefits relative to relevant electronic systems in terms of a central interrogation system for various types of sensors, signal quality and measurement noise immunity, sensor distribution and

Linear Bragg grating

Chirped Bragg grating

FP cavity grating

Long-period grating

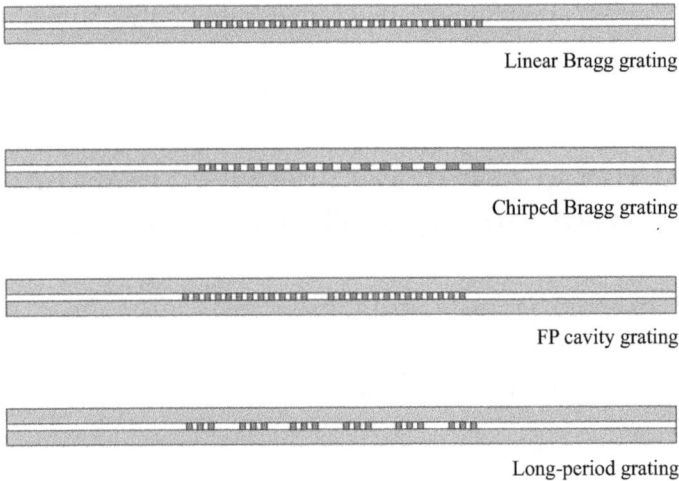

Figure 8.1   Schematic of different types of FBG

multiplexing that enables remote sensor placement and a high-sensor channel capacity. The advantages of a fibre-optic solution include [1–5]

- integrated structures on an optical fibre or waveguide for long-term alignment stability;
- high signal integrity due to insensitivity to EMI/radio frequency interferences;
- freedom from sparking and electrostatic discharge (ESD);
- lightweight, flexible fibre-optic harness for significant mass savings with high signal integrity;
- low-loss (<1 dB/km) fibre-optic bidirectional optical signal routing;
- efficient WDM for high sensor capacity per fibre;
- low power requirements per sensor or device (mW);
- signal interrogation based on a change in the optical wavelength for high noise immunity and
- remote compact central interrogation system (<1.5 kg) based on a tunable fibre laser.

Fibre optics are especially advantageous for systems operating in harsh environments such as spacecraft planetary landers and rovers, as well as critical terrestrial systems (underwater operations, mining, chemical processing and nuclear systems).

An FBG [3] can be serially directly written onto a segment of an optical fibre to provide a tailored spectral reflectance over some range of optical wavelengths, typically using a ultraviolet (UV)-pulsed laser and a suitable photolithographic mask to spatially modulate the refractive index (RI) of the fibre core. The FBG can be used in reflectance or transmittance modes. Various FBG structures are possible to imprint in the fibre core to facilitate a wide palette of selectable spectral and spatial optical characteristics:

- uniform-period Bragg reflector for high reflectance at a selected wavelength;
- chirped-variable period Bragg reflector for reflectance over a range of wavelengths and at different positions along the grating;
- FP-type grating and
- long-period Bragg reflector for broad spectral reflectance.

The centre wavelength (CWL) of an optical FBG can be related to changes in the ambient temperature and external stress by [4]

$$\frac{\Delta \lambda_B}{\lambda_B} = (1 - P_e)\frac{\Delta L_F}{L_F} + (\alpha_F + \varepsilon_T)\Delta T \tag{8.1}$$

$\Delta \lambda_B$ is the delta λ_B, i.e. the variation in the Bragg period; P_e is photoelastic coefficient of the optical fibre; ΔL_F is the delta L_F, i.e. the variation of the length of the fibre; ε_T is the thermo-optic coefficient; α_F is the linear thermal expansion coefficient of the fibre; ΔT is the delta T, i.e. the variation of the ambient temperature.

Where $P_e = 0.22$ is the photoelastic coefficient of the optical fibre, $\alpha_F = 0.55 \times 10^{-6} \text{ K}^{-1}$ is the linear thermal expansion coefficient of the fibre, $\varepsilon_T = 8.31 \times 10^{-6} \text{ K}^{-1}$ is the thermo-optic coefficient and L_F is the length of the fibre.

The optical fibres with FBG-sensing elements can be specially mounted in a miniature package using MEMS techniques to enhance their sensitivity to the desired physical property and provide some environmental protection.

The change of the physical measurand changes the spectral characteristics of the FBG-based sensor. As a result, the sensor measurement is independent of the absolute intensity of the optical signal allowing long-term stable sensor calibration.

8.2   Microelectromechanical systems/fibre Bragg grating temperature sensing

The FBG sensor temperature is obtained from an estimate of its CWL. Typically each optical fibre can accommodate several such FBG sensors along its length, each operating over a different spectral wavelength range.

The FBG wavelength can be interrogated by scanning a tunable laser across the reflectance spectrum of the FBG sensor. This output is reflected by the FBG based on its spectral reflectance characteristics. This results in a cross-correlation of the spectral output intensity of the laser source with the reflectance spectrum of the FBG sensor. A photodetector measures the integrated intensity of the cross-correlation of the laser output and the spectral reflectance characteristic of the FBG sensor.

The scanning tunable laser interrogation technique provides a basic wavelength accuracy of about ±0.001 nm for the CWL based on peak detection. This will yield a nominal temperature measurement accuracy of about 0.05°C using simple peak detection techniques and the scanning tunable laser interrogation [5].

One of the challenges with using FBG for temperature measurements is that the grating CWL is sensitive to both temperature and strain. A special proprietary packaging was developed that nearly triples the effective sensor sensitivity to temperature ($\Delta\lambda/\Delta T \cong 0.03$ nm/°C) (Figure 8.2), relative to the sensitivity of the bare FBG (about 0.013 nm/°C) for improved measurement accuracy. The special packaging also decouples the FBG from the sensor mounting and surface strain. This enables good thermal contact while maintaining the FBG sensor calibration.

Figure 8.3 shows a cabled FBG T-sensor line for the fibre sensor demonstrator (FSD) flight unit with four packaged FBG sensors. Each of the sensors operates over a different band of optical wavelengths. A spectral bandwidth of 5 nm was

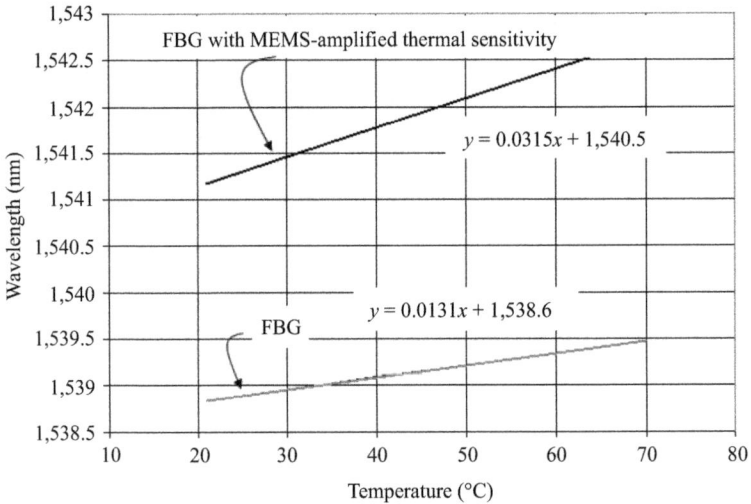

Figure 8.2   Experimental measured temperature response of bare FBG and packaged FBG with MEMS-amplified thermal sensitivity

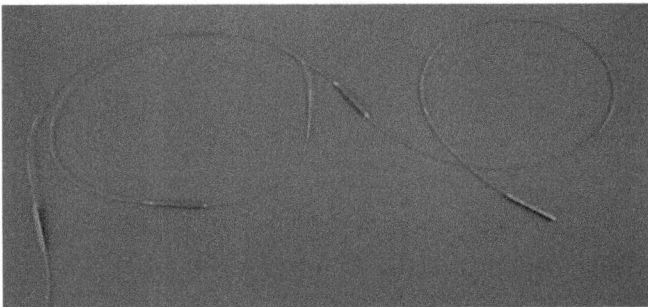

Figure 8.3   Photograph of a packaged FBG T-sensor line with protective cabling and four inline packaged FBG/MEMS temperature sensors (courtesy of MPB Communications Inc.)

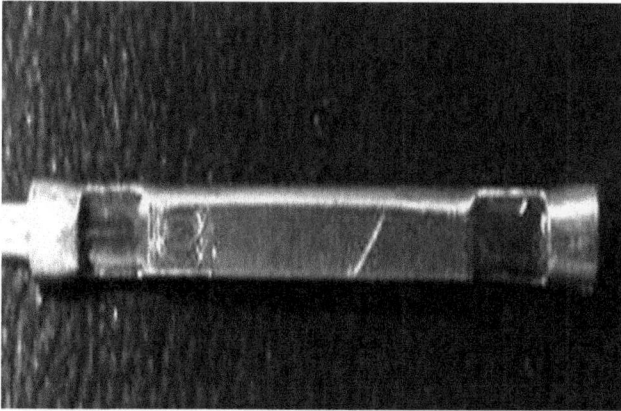

Figure 8.4   Photograph of the inner T-sensor assembly using SS microtubing and additional strain relief at both ends

provided for each sensor to avoid any possibility of a spectral overlap between adjacent sensors under worst case conditions.

Figure 8.4 shows the internal assembly of the FBG within selected stainless steel (SS) microtubing. This is then inserted into a thermally conductive outer jacket such that mounting strain at the outer jacket is not coupled to the inner FBG assembly, allowing the decoupling of strain and temperature.

8.3   Tunable fibre-optic variable delay line

A linearly chirped FBG (Figure 8.5(a)) can be used to provide a wavelength-dependent time delay in the frequency components of a reflected optical pulse. The path length of the reflected signal varies with its wavelength due to the FBG chirp. One application of this is to compensate for the chromatic dispersion introduced by the carrier fibre and other network components in fibre-optic communications networks (Figure 8.5(b)). Figure 8.6 is a plot of the delay in picoseconds as a function of the FBG wavelength for a 10-cm-long linearly chirped FBG.

By integrating the chirped FBG with an actuator, the time delay at a given spectral wavelength can be dynamically tuned. This can be accomplished thermally by inducing a thermal gradient across the FBG, or mechanically by compressing and/or stretching the FBG.

The dispersion compensation unit (DCU) employs a linearly chirped FBG to provide a wavelength-dependent time delay in the frequency components of the reflected pulse that can compensate for the chromatic dispersion introduced by the carrier fibre and other network components. As the optical signal travels through the chirped FBG, different wavelength components are reflected back at different positions along the grating. The maximum time delay depends on the length of the grating and the RI of the fibre. A long fibre grating (3–6 cm) is used to facilitate

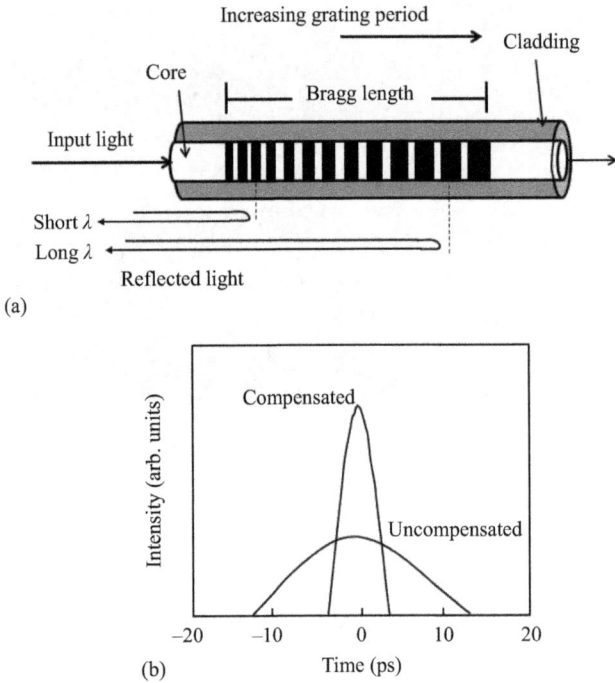

Figure 8.5   (a) Linearly chirped FBG as a wavelength-dependent variable time delay and (b) optical pulse dispersion compensation

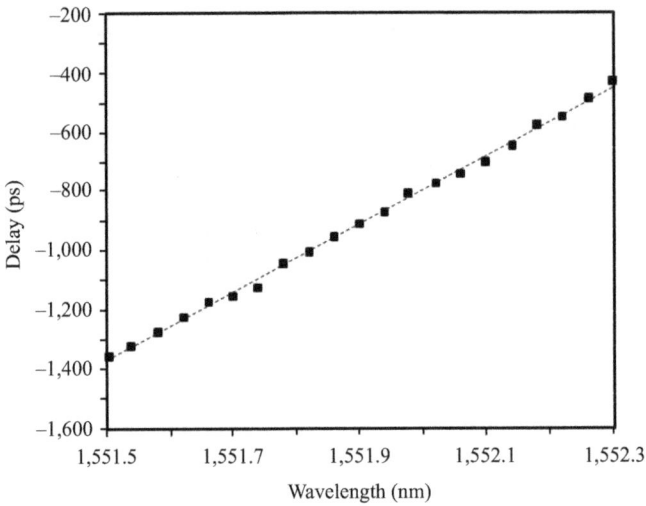

Figure 8.6   10-cm-long linearly chirped FBG for variable time delay versus wavelength

relatively large dispersion compensation. A robust thermomechanical actuator provides dynamic tuning of the grating chirp and CWL. The DCU can compensate between 500 and about 2,000 ps/nm. Figures 8.7 and 8.8 display typical examples of a photo of a DCU prototype with the control electronics for the FBG wavelength tuning and an actuator based on FBG sensors for thermal gradient control, respectively.

The DCU effectiveness was tested using a link span of 335 km of single-mode fibre (SMF). The test system was run at STM-16 data rates (2.5 Gbits/s). The bit error was monitored on the STM-16 Analyzer. The results without and with the DCU are summarised in Table 8.1. With the DCU, the bit error rate decreased by a factor of about 40.

Figure 8.7 *FBG actuator assembly for thermal gradient control across the FBG*

Figure 8.8 *Photograph of prototype DCU with the control electronics for the FBG wavelength tuning [142 (L) × 32 (W) × 100 mm (H)]*

Table 8.1 *Summary of DCU testing in a 335-km SMF telecom test set-up at STM-16 data rates (2.5 Gbits/s)*

Device under test	Bit error rate	Bit error count
No compensation replaced by a 10 dB fix attenuator	7.85×10^{-9}	5,644
1,600 ps/nm of DCU	2.06×10^{-10}	148

8.4   Microelectromechanical systems/fibre Bragg grating pressure sensor

FBG sensors can also be applied to pressure measurements in critical environments such as monitoring the amount of propellant in a spacecraft propulsion subsystem, where the added safety due to freedom from ESD and sparking is of great benefit.

A combined pressure/temperature (P/T) sensor with a thermal reference wavelength FBG was developed for the FSD [6], currently flying and still operating on ESA's PROBA-2 spacecraft. This contains three FBG sensors in series written on the same fibre, each operating over a different wavelength range. One of the sensors is thermomechanically amplified to maximise sensitivity to temperature of the P/T sensor body temperature, while minimising sensitivity to pressure or stress. The second FBG sensor is mechanically linked to a membrane actuator (Figure 8.9(a)). The membrane hermetically isolates the pressure-sensing FBG from the propellant being sensed. A third sensor is a wavelength reference FBG that is located within the remote central interrogation unit. This is used to provide a reference absolute wavelength for the sensor line. It employs a thermal sensor packaging to minimise the sensor CWL variation with temperature [5].

The P/T sensor high pressure cavity that is exposed to the propellant employs an electropolished, unibody SS construction that avoids any welds or epoxy for unprecedented reliability, leak tightness and cleanliness (Figure 8.9). The thin SS membrane is micromachined directly into the cavity and forms the mechanical interface between the propellant gas and/or liquid and the pressure-sensing FBG. Deflection of the SS membrane in response to the propellant pressure acts to compress the sensing FBG. The pressure-sensing FBG operates in compression within a close-tolerance capillary tube to maximise the fibre strength and the attainable wavelength shift with pressure (up to 40 nm in compression). A small SS piston links the membrane deflection to the fibre compression. The pressure readings are independent of the gas composition.

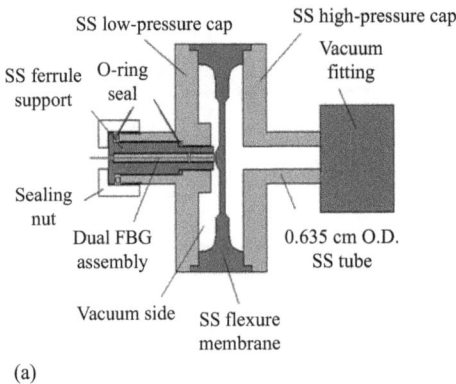

Figure 8.9   (a) Schematic of the P/T sensor with SS membrane and (b) photograph of the mechanical assembly (OD, outer diameter)

The second integral temperature-sensing MEMS/FBG structure is employed to compensate the pressure readings for any sensor temperature fluctuations.

The achieved P-sensor response to the applied pressure is very linear with minimal hysteresis comparable to the reference pressure measurement accuracy. Figure 8.10 shows the response of the P/T pressure and temperature sensing using FBG. The temperature-compensated peak hysteresis was about ±0.015 nm relative to a full scale FBG CWL shift of about 12.75 nm for 600 psi applied pressure. The T-sensor response is totally isolated from the pressure effects due to the special mounting that was employed, showing no variation in its CWL with the applied pressure.

Figure 8.11 shows the P/T sensor mounted to the SSTL propulsion system of PROBA-2 to monitor the pressure with the propellant tank.

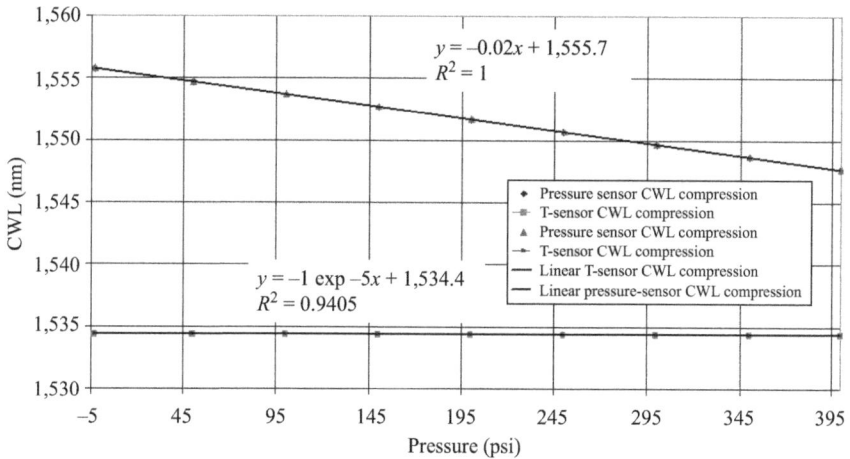

Figure 8.10   Response of the P/T pressure- and temperature-sensing FBG to the applied pressure

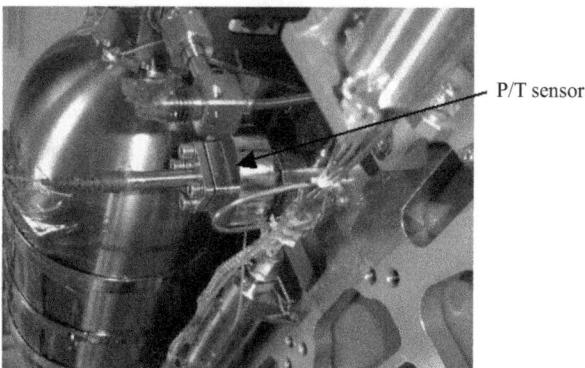

Figure 8.11   Photograph of the FSD flight P/T sensor integrated with the SSTL propulsion system for PROBA-2

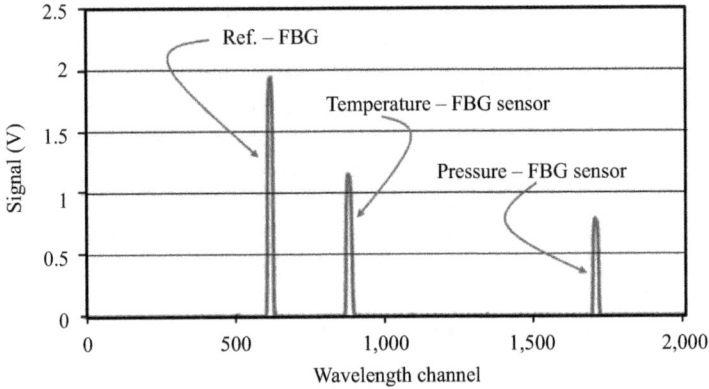

Figure 8.12   On space orbit read-out of the FSD P/T sensor line

Figure 8.12 shows a measured scan of the FSD P/T sensor line with integral a thermal reference FBG, as downloaded from the ESA PROBA-2 spacecraft in orbit. Readings from all three sensors were obtained. The signal quality provided on orbit by the FSD fibre optics was very good, similar to that during the ground testing and sensor calibrations.

8.5   Inertial sensing

Inertial sensing, such as acceleration, is widely used in transportation, including airbag deployment mechanisms on vehicles. It is also very critical in space for launchers, spacecraft, landers and planetary rovers to provide some autonomy in the relevant navigation systems and monitoring of vibration levels during various manoeuvres.

Current inertial measurement units are based on integrated-fibre-optic gyroscopes for rotation sensing and MEMS capacitive accelerometers for acceleration sensing. These can meet the measurement performance requirements for various space missions; however, they have limited reliability, with a mean-time to failure of about 3 years, require special mounting for vibration/shock dampening, and the associated high power (12 W/unit) and cost (>10^5/unit) exceeds the typical budget allowance for small spacecraft. Moreover, typical launchers employ dozens of accelerometers that require a bulky shielded signal harness. The penalty is additional mass and fuel for the launch.

This section provides an overview of the μNAV^{TM} micronavigator inertial acceleration sensing. This is based on miniaturised photonic sensors integrated on a silicon-on-insulator (SOI) microphotonic microchip, similar to an electronic integrated circuit. In-plane mirrors based on photonic bandgap (PBG) principles are integrated with MEMS actuators and proof-mass (Mproof) to facilitate the conversion of acceleration and rotation to a relevant change in the intensity or wavelength of an optical signal.

The non-contact optical interrogation of the PBG/MEMS inertial sensors provides a true linear sensor response with over 60 dB of measurement range, significantly improving the performance relative to traditional MEMS electronic capacitance sensors. The acceleration or rotation is converted into a corresponding linear change in the optical signal wavelength (FP etalon sensor) or intensity [variable optical attenuator (VOA) sensor] that can then be interrogated very precisely using an optical signal.

We focus here on the VOA-sensing junction approach. This requires both a reference optical channel for monitoring the reference optical signal intensity and a sense channel to interrogate the sensing junction.

Semiconductor electronics started with basic devices such as transistors, diodes, resistors fabricated on silicon wafers that were then diced and assembled in individual packages with gold wire-bonded interfaces to the external circuit board and the required I/O and bias electrical signals. In the 1970s, this advanced to several devices integrated on a common silicon die to provide a basic integrated function, such as a digital gate (NAND and so on) or a resistor divider network, or for analog signals, an operational amplifier. The level of integration advanced in the 1980s for the integration of 10s and 100s of devices to provide greater functionality such as memory chips, programmable logic arrays, multiplexers and analog/digital converters. In the 1990s, the electrical device area was further scaled down and full electrical systems on a microchip were realised to facilitate the personal computer revolution. To facilitate this required two basic ingredients, the basic functional blocks and the associated microfabrication recipes and equipment. Today, semiconductor technologies and associated products comprise a multi-trillion dollar industry that impacts every aspect of our lives.

MPB Communications Inc. is taking a similar road-map to the implementation of PBG/MEMS microphotonics, focusing on the needs for space where high reliability and high functional densities per unit mass are key requirements and technology drivers. The selected platform is SOI as this allows the adaptation of the relevant microfabrication processes that have been previously developed for semiconductor electronics. The added technology from MEMS is deep reactive ion etching (DRIE) to facilitate the two-dimensional (2D) and three-dimensional MEMS micromechanical structures that can respond to acceleration and/or rotation.

The approach has entailed a systematic development and validation of the required technology building block and associated microfabrication. The various generic building blocks are subsequently prototyped and optimised. The resulting library of generic functional component blocks can be subsequently integrated together to enable various, more complex multifunctional microphotonic MEMS/PBG microchips. As the component library expands, more sophisticated microphotonic/MEMS systems can be realised on a SOI microchip at minimal additional development cost.

Several of the basic required components were prototyped and experimentally validated in 75-μm thick silicon device layers using SOI technologies, including

1. multimode channel waveguides
2. total internal reflection T-bar 1:2 optical signal splitter

3. tapered output optical signal waveguide collimator
4. planar micromachined PBG mirrors
5. MEMS silicon flexure springs and Mproof inertial sensing
6. MEMS electrostatic comb drive
7. advanced input/output U-grooves to facilitate robust fibre integration for interfacing to the external world

A schematic of an integrated optic linear accelerometer based on VOA of the transmitted optical signal is shown in Figure 8.13. An ungroove at the optical input facilitates the integration of a standard 125 μm outer diameter cladding fibre to provide the sensor input signal. Passive 1:2 T-bar optical signal splitters provide four signal channels; two are used for reference signal intensities (Io/4 and Io/8, where 'Io' is the input signal intensity) and two for the differential VOA sensing.

The acceleration sensing is provided by a movable MEMS Mproof that is suspended using a set of silicon flexure springs linked to anchors. The flexure springs are designed to provide a selected spring constant Ky along the y-axis and be relatively still along the x- and y-axes ($Kz, Kx > 100Ky$).

The resultant deflection of the MEMS Mproof, $\Delta y(acc)$, is proportional to the applied acceleration acc:

$$Facc = Mproof \times acc = Ky \times \Delta y(acc) \tag{8.2}$$

Facc is the displacement of MEMS Mproof as a function of the acceleration; $\Delta y(acc)$ is the resultant deflection of MEMS Mproof as a function of the acceleration.

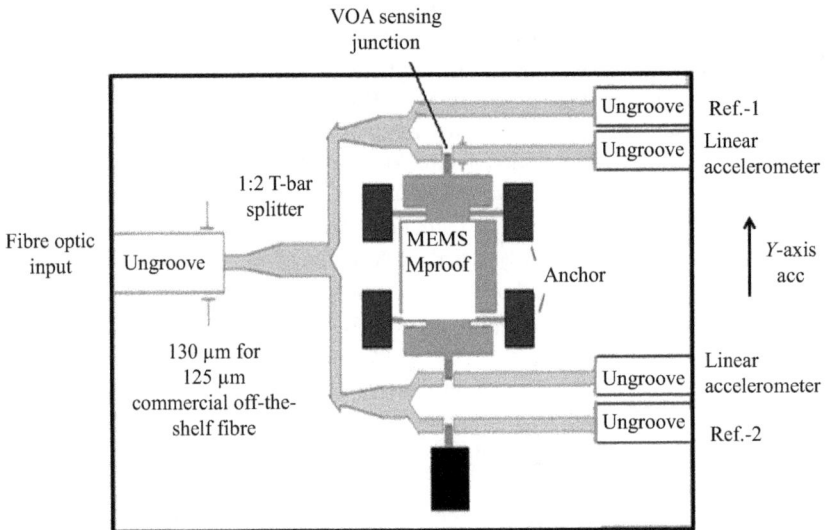

Figure 8.13   Advanced VOA accelerometer design using a differential VOA sensor structure with integrated sense and reference waveguide channels

This displacement positions a PBG mirror structure, micromachined onto the ends of the Mproof, between the sensing channel waveguides. This acts as a shutter to reduce the optical coupling between the two waveguides, resulting in a transmitted optical intensity that is proportional to the applied *y*-axis acceleration.

Ansys Multiphysics^{TM} [6] and Coventor^{TM} [7] were used to assist the design of the MEMS Mproof and the supporting silicon flexure springs. An example of a MEMS/flexure simulation is given in Figure 8.15. This facilitated optimisation of the silicon flexures in terms of the selected spring constant along the desired sensing direction and high stiffness in the orthogonal directions.

The integrated-optic device 2D patterning on SOI requires a relatively dry DRIE of the silicon device layer. The microfabrication process optimisation was performed using the DRIE facility at McGill University in collaboration with Ecole Polytechnique (Canada). As shown in Figure 8.14, a 75 μm etch depth in the silicon device layer was achieved with excellent verticality and relatively good sidewall smoothness with respect to the requirements for the optical channel waveguides. This allows the interfacing of standard, highly robust telecom optical fibre to the PBG/MEMS microphotonic sensor integrated-optic microchips.

Figure 8.15 is a scanning electron micrograph of a patterned VOA-sensing junction. Perforations in the MEMS silicon layer were used to facilitate the MEMS lift-off by under etching the intermediate silicon dioxide sacrificial layer. Anchors and bridges are used to prevent the ends of the channel waveguide from lifting up. The VOA shutter is inserted into the gap in the channel waveguide in response to an applied acceleration to convert the applied acceleration or tilt into a variation in the magnitude of the optical signal that is transmitted to the mating channel waveguide.

Single-mode input and output fibres were bonded to the U-grooves using UV-curing optical adhesives. Light at 1,550 nm from a 4 mW laser source was injected into the input channel waveguide *via* the integrated-input optical fibre.

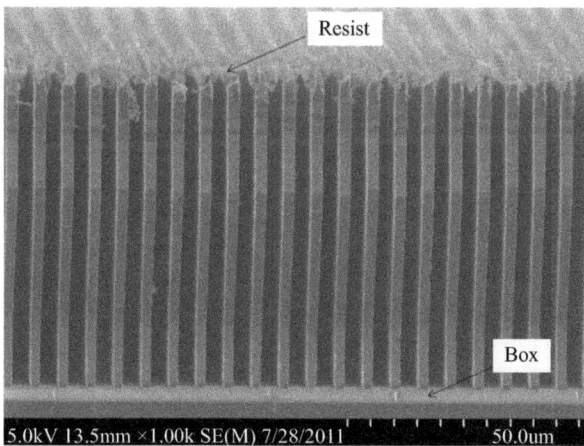

Figure 8.14 Scanning electron microscopy (SEM) edge view of 75 μm deep silicon etching of the test structures

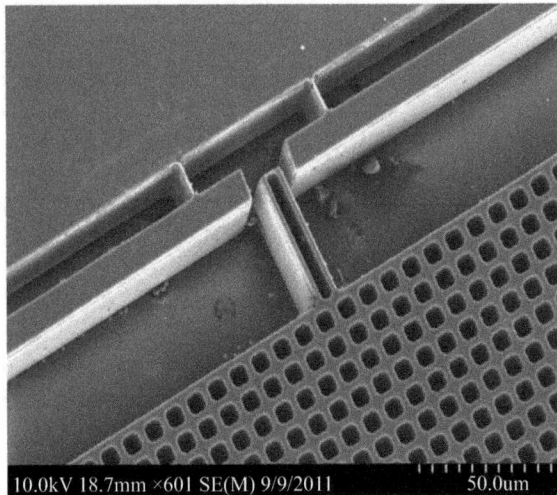

Figure 8.15   SEM micrograph example of a VOA-sensing junction with anchors to reduce channel waveguide lift-off

The transmitted light was collected by a second optical fibre coupled to the output U-groove. The resultant output signal was detected using an InGaA photodiode.

By using a multimode output fibre or close coupling the detectors at the output of the channel waveguides, much higher overall optical throughput is feasible to further increase the usable measurement dynamic range and signal-to-noise ratio (SNR). With the multimode output fibre coupling, much less gain transimpedance is required for the same voltage signal with about 21 times better net optical signal coupling than a single-mode output fibre.

Static acceleration measurements were accomplished by mounting the PBG/MEMS fibre-pigtailed accelerometer on a tilt table. The tilt angle was varied to vary the effecting force due to gravity acting along the sensing direction.

As summarised in Figure 8.16, the output optical power decreased as the angle of inclination of the accelerometer was increased since more light was blocked by the Bragg shutter moving progressively into the gap between the channels waveguides in the VOA-sensing junction *via* the MEMS actuator response to the applied tilt.

The sensor response to the acceleration was approximately linear. Sensitivity to lower accelerations can be improved by using a bigger Mproof and/or softer springs at the expense of a lower full-scale range. By using two parallel accelerometers with different Mp/Ky, we can provide measurements over a wide range of accelerations. The results are summarised in Table 8.2 for two different VOA accelerometers with different MEMS Mproof.

A Bruel & Kjaer Mini-Shaker was used to evaluate the optical performance of the device during and after undergoing some sinusoidal vibrations. The integrated-VOA sensor with input and output fibres has been mounted on the shaker and

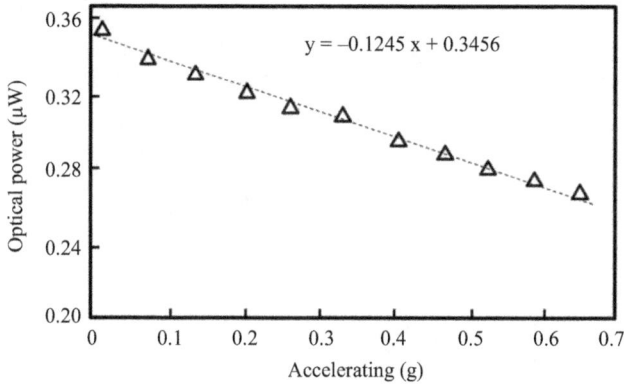

Figure 8.16   Measured 2D VOA sensor response (1,000 × 1,000 × 75-μm Mproof, with 12-μm wide VOA-sensing junction)

Table 8.2   Summary of the PBG/MEMS VOA accelerometer results

Accelerator	$\Delta V/V_{ref}$ for 1 g (ΔV: voltage variation, V_{ref}: reference voltage)	V_{ref}	Silicon MEMS proof mass size (μm^3)	Responsivity ($\Delta V/g$)	Resolution at SNR $= 1$
No. 1	0.055	6.8	720 × 720 × 75	0.374	Current noise level of 13 μV 35 μg
No. 2	0.04	6.8	720 × 720 × 75	0.272	50 μg

underwent vibrations from 1 to 100 Hz. The testing indicated that the sensors could follow vibrations out to above 1 kHz.

The test device consists of a VOA 2D accelerometer with integral input and output optical fibres mounted on an aluminium base, as shown in Figure 8.17.

The integrated linear accelerometer was able to follow the applied sinusoidal shaker vibrations to above 100 Hz, as summarised in Figures 8.18. The preliminary testing indicated that the accelerometer response, with current electronics, can exceed 1 kHz.

8.6   Conclusions

This chapter reviewed the integration of fibre optics with FBG with MEMS actuators to facilitate various tunable fibre optic and integrated-optic devices for optical signal processing, such as variable time delay lines, and the sensing of physical parameters, such as temperature, pressure and acceleration/tilt.

*Figure 8.17   Assembled VOA-based accelerometer mounted on small Bruel &
Kjaer Mini-Shaker*

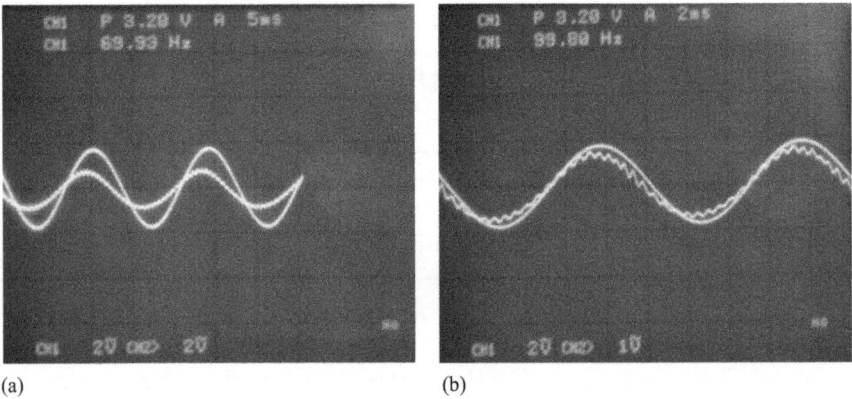

(a)   (b)

*Figure 8.18   Oscilloscope pictures of the linear accelerometer alternating current
(AC) response at different applied sinusoidal vibration frequencies
(a) 69.93 and (b) 99.9 Hz, showing the AC source signal applied to
the shaker and the associated measured accelerometer response*

Through the FSD flight validations on ESA's PROBA-2 spacecraft, as launched in November 2009 and still in operation in 2016, this technology can provide
reliable long-term operations in harsh environments with significant benefits relative to relevant electronic sensors and signal processing in terms of immunity to
EMI/electromagnetic compatibility radiation, low-loss, and lightweight fibre-optic
signal harness, sensor mass and signal quality.

A remote central interrogation system can be employed to operate the various
censors with a relatively high-sensor channel capacity through WDM on each fibre.

References

[1]   I. Mckenzie and N. Karafolas, *Proceedings of the SPIE*, 2005, 5855, doi:1117/12.623988.

[2]   M. Ott in *Technology Validation of Optical Fibre Cables for Space Flight Environments*, Sigma Research and Engineering, NASA Goddard Space Flight Center, Greenbelt, MD, USA, 2000. *https://pdfs.semanticscholar.org/ 285f/b95729f846791c68153b523f9e30911e393b.pdf.*

[3]   K. Hill, Y. Fujii, D. Johnson and B. Kawasaki, *Applied Physics Letters*, 1978, **32**, 10, 647.

[4]   R. Kruzelecky, J. Zou, N. Mohammed, *et al.*, *Proceeding of the 6th International Conference on Space Optics*, ESA SP-621, ESTEC, Noordwijk, The Netherlands, 27–30 June 2006.

[5]   R. Kruzelecky, J. Zou, E. Haddad, *et al.* in *Fiber-optic Sensor Demonstrator (FSD) for the Monitoring of Spacecraft Subsystems on ESA's PROBA-2*, ISCO 2010, Toulouse, France, 2010.

[6]   Ansys Multi Physics Software. *https://www.ansys.com/products/platform/ multiphysics-simulation.*

[7]   Coventor Software – MEMS design and simulation software suite. *https:// www.coventor.com/mems-solutions/products/coventorware/.*

Chapter 9

Summary and challenges of the fibre optic sensor technology

Over the past six decades, optical fibre sensing (OFS) has been developed to be employed in testing and improving the health integrity, efficiency, safety and reliability of various structures, different vehicles, electronic devices, medical equipment and much more over a myriad of industries. Recent developments have capacitated fibre optic sensors (FOS) to bolster its capabilities to integrate new fields across applications in medical, energy and space. This is enabling engineers deal with issues they are facing with innovative designs.

The real-world opportunities of the OFS technology applications are just unlimited, as well as huge dimensions of possibilities for the future. Basically, fibre Bragg gratings (FBG) is a key element in the exponentially growing up optical fields of communications and sensing. Even though its demonstrated efficiency, the FBG device is rather simple. In its simplest form, a FBG is expressed by a regular and periodic modulation of the refractive index within a single-mode optical fibre core, where the fronts of the phase propagation are perpendicular to the longitudinal direction of the fibre and with grating planes having a constant period. Optical signal (light wave), navigated along the core of an optical fibre, is scattered by each grating plane. If the Bragg condition is met, the contributions of reflected optical signal from each grating plane will add in a constructive way in the reverse direction to form a back-reflected peak with centre wavelength defined by the grating period.

9.1 Summary of the book

This book addresses the critical challenge of developing FBG for applications that require operation in harsh environments and focuses on space-frontiers application. It covered all the aspects of the technology, i.e. from basic research through design, fabrication and testing to the industrial implementation of high-temperature and radiation-resistant optical fibres. The areas of testing encompassed high-temperature annealing and effect of the exposure to gamma radiation encountered by satellites.

The book starts with providing a detailed introduction to the key element of the FOS, namely FBG that includes the fundamental understanding of the spectral properties of the fibres used. Then, some of the most recent developments in the application of FBG sensors for extreme environment conditions are reviewed.

Although the dominant technologies used today to perform temperature measurements still are based on electrical sensors, optical sensors were shown to offer a promising alternative to challenging applications.

Harsh environments, distributed systems, space and long-term deployments are typical examples where the characteristics of the FBG systems can make provide the clear advantage and a highly effective solution as compared to conventional electrical sensors.

The introduction is followed by a review of the Bragg gratings technologies in the optical fibres, where the photosensitivity in the optical fibres was introduced and detailed. Then, the various specificities and properties of the Bragg gratings are examined and the most important advancements in the devices conception and various applications are presented. The most dominant fabrication technologies including phase mask, interferometry and point-by-point are described in detail with appropriate schematics, figures and with reference to their advantages and disadvantages in utilising them for writing the Bragg gratings.

The last chapters of the book are oriented toward space applications. Chapter 6 highlights deals with the presence in the space of micrometeoroids and orbital debris that are characterising the lower Earth orbit (LEO), which in turns presents a true hazard to orbiting satellites. We addressed a short review of the space debris challenges and reported on the evaluation of FBG sensors in the LEO environment. Then the most recent experimental results are presented on the applications of FBG with self-healing composite materials used in space. In Chapter 7, we have presented the fibre sensor demonstrator flight validations on European Space Agencys' (ESA) PROBA-2 spacecraft, as launched in November 2009 and still in operation until now. This technology has proven indeed to provide reliable long-term operations in harsh environments with significant benefits relative to relevant electronic sensors and signal processing in terms of immunity to electromagnetic interference/electromagnetic compatibility radiation, low loss and lightweight fibre optic signal harness, sensor mass and signal quality.

Finally, in Chapter 8, we reviewed the integration of fibre optics with FBG with microelectromechanical systems actuators to facilitate various tunable fibre optic and integrated optic devices for optical signal processing, such as variable time delay lines, and the sensing of physical parameters, such as temperature, pressure and acceleration/tilt.

9.2   Recent tendency and challenges of the fibre optic sensing technology in space and composite structures

The improvements of OFS technology in terms of sensing sensitivity, spatial resolution, data rate and its advancements in general not only boost the capacity of multiple industries to fix complicated issues, and help qualified engineers to progress beyond the issues they are facing today, but also to innovate and outlook into the future.

This technology continues to grow and advance, with a direct impact on architecture and sophisticate configuration across many fields, including medicine, environment, energy and aerospace in the light to face challenges and fix problems that do not exist yet.

OFS technology is adaptable to various scenarios. It is a flexible technology that could be either integrated as a platform designs and/or component of critical systems for which the real-time monitoring is needed or stand alone as an advanced testing process.

The adoption of FOS in space application over mature and available alternatives is often motivated by the two following factors:

- Enhanced performance (they are indeed improved when FOS is employed individually).
- Their unprecedented multiplexing capability.

In some particular space missions, where there is no radiation prerequisite, the competitive cost of FOS adds a net advantage. Various question still remain open, such as what is the most convenient integration techniques for FOS under harsh thermal and mechanical extreme conditions.

On the other hand, the effect of the microgravity environment on the sensing integrity should be taken into account to ensure an accurate measurement [1]. Indeed, at high-temperature environment, especially above 926.85°C, measuring both the temperature and pressure become really challenging. In addition, the sensitivity to small hydrogen leakage and determining the fuel level under microgravity conditions remain open challenges. The OFS technology could be also of a key role for the space transportation reusable vehicle that necessitates complex health monitoring. The interest behind the involvement of a committed spacecraft-sensing subsystem by means of hybrid integration between the FOS and any other established technology is still to be investigated and validated [2]. In fact, to adopt such a hybrid technology, the spacecraft manufacturers need first to necessary acquire a mature experience in implementing FOS system into the existing space structures and gain the needed confidence and comfort in its use. As a matter of facts, the applications of OFS technology considered by the ESA are summarised in Table 9.1.

On the other hand, integrating FOS within composite material is a minimally invasive process; however, for industry, numerous issues are still under investigation.

One of the main challenges is the development of a steady procedure to connect the optical sensor [3,4]. Custom-created connectors are promising techniques; however, these optical fibres when they are equipped with connectors may change the viscosity of the composite which become somehow brittle especially at the edges of the structure [5]. In addition, after installing the connectors into the fibres, trimming the edges of the composite become highly challenging. Alternatively, the free-space coupling technology is much more suitable than surface and edge mounted connectors [6]. As a matter of fact, an original process of free-space passive coupling of the optical signal into the FBG sensors was investigated and consisted of a 45° angled mirror that is combined with the FOS and integrated into the optical fibre [7].

Moreover, to fix the problems related to the connectors and optical fibre edges, a novel technique based on the reconnection was proposed and consisted of settling the connectors on a six-axis automatic stage along with the imaging support of charged coupled device camera [8].

Table 9.1   FOS applications by the European Space Agency

Satellite subsystem	Sensor
Structure and payload	Temperature, static and dynamic strain
Propulsion	Temperature, pressure, valve status, remaining liquid propellant (microgravity)
Attitude control	Rotation, acceleration
Launcher subsystem	
Propulsion	Temperature, pressure, acceleration, leak detection, valve status
Structure	Temperature, strain, acceleration, leak detection
Atmospheric entry vehicles	
Thermal protection system	High temperature
ISS	
Experiment	High temperature (up to 2,000°C)
Ground testing	
Rocket nozzle	Pressure, thermal flux (high temperature)
Antennas reflectors	Strain (dynamic in high frequency)
Solar sails	Strain, temperature

Additionally, as the FOS and/or the optical fibres diameter is generally larger than that of the reinforcement fibres, structural damage could occur in the composite material. To overcome this limitation, researchers are developing new shape with a reduced optical fibre diameter having an adapted coating. As a matter of fact, the draw tower grating fibre is one of the typical examples of such a new class of thinner optical fibres, where the diameter is about 80 μm rending these thinner FOS less invasive when integrated within a carbon-fibre-reinforced laminate composite, especially when used for space application.

A traditional FBG sensor's essential response is to an axial strain; however, the transverse strain has definitely some impact as well [7,9]. Nonetheless, it is somehow problematic to dissociate from the FBG spectral signal the axial from the transverse strain.

For that, in applications requiring strain mapping, especially when dealing with structural health monitoring including the cracks detection, delamination and damages, often the multiaxial strain measurement is employed. This in turn highlights the relevance of developing a new optical-sensing procedure that furnishes both axial and transverse strain measurements at the same time. A literature survey shows us that when FBG are inscribed in the highly birefringent (HB) fibres and/or HB microstructured fibres (MOF), measuring the transverse and axial strain simultaneously becomes possible [10,11]. In fact, an FBG fabricated in an HB fibre could display two distinguished Bragg peaks, corresponding to the two orthogonally polarised modes. The variation of the Bragg peak separation is subject to the phase modal birefringence change that is induced either by a given transverse load and/or temperature variation. The sensitivity of MOF to various factors (strain, pressure, radiation, temperature and so on) is dictated by the type of the optical fibre employed [12]. Table 9.2 summarises the main advantages/disadvantages and applications of FOS.

Table 9.2 Comparison of FOS technologies

FOS technology	Advantages	Disadvantages	Remarks main	Applications
Standard FBGs	Most accepted technology, allows for point measurements of strain and temperature	Temperature and strain cross-sensitivity issues	Typical strain sensitivity ~1.2 pm/μɛ and typical temperature sensitivity ~11.6 pm/°C	Strain, temperature, vibration, cure process, localised damage, etc.
FBGs written in MOF	Can discriminate both axial and transverse strain components of composite material with insignificant temperature sensitivity	FBGs written in bow-tie fibres have temperature and strain cross-sensitivity. But FBGs written in microstructured fibre (MOF) have lower strain sensitivity compared to FBGs written in bow-tie fibres	The cross-sensitivity issue can be resolved by using FBGs written in low-temperature-sensitive MOFs	Multidirectional strain sensing, localised damage, etc.
Interferometric fibre optic sensors	Possesses higher temperature and strain sensitivities and are flexible in terms of size	Temperature and strain cross-sensitivity issue, and brittle sensor	The cross-sensitivity issue can be resolved by using low-temperature-sensitive MOFs	Strain, temperature, vibration, cure process, localised damage, etc.
Polarimetric sensors	Sensitivity can be tuned by choosing different optical fibre types and sensor lengths	Difficult to measure strain/temperature at localised points provide information averaged over the sensor's length	The cross-sensitivity issue can be resolved by using low-temperature-sensitive HB-photonic crystal fibre	Strain, temperature, vibration, cure process, etc.
Fibre optic microbend sensors	Can measure continuous strain profile in a composite material using single optical fibre	Low accuracy	Output signal is strongly attenuated by any mechanical wave propagating in the composite material	Delamination and damage detection
Distributed sensors	Can measure continuous strain/temperature profile in a composite material using single optical fibre	For better resolution require the use of spectral demodulation techniques that are expensive and bulky	Appropriate sensing technology can be selected based on the application and its requirements	Strain, temperature, delamination, damage detection
Hybrid sensors	Two or more FOS operate in a combined manner to eliminate the disadvantages of individual FOSs, providing accurate and independent strain/temperature information	Since two or more sensors are employed, complicated interrogation methods are needed	Capable of discriminating between strain, temperature and thermal strain	Strain, thermal strain, temperature, vibration, cure process, damage point, etc.

Additional concern related to inserting the FOS within the composite part for applications where the weight (e.g. in space) is a sensitive factor is the large size of traditional FOS interrogators, which could present serious issues for sensing of the composite parts that are, for example, in a continual motion, such as rotors, turbines and so on [13].

To a certain limit, this problem could be overcome by miniaturising the interrogation systems using either flexible polymer waveguides or photonic integrated circuits [14,15]. This kind of flexible interrogators surface permit the combination of arrays of photodetector with the wireless communication technology and hence open the way to smart sensing of composite parts either in continual motion. It is worth noting at this level that the FOS embedding process is rather a hard and demanding task. Thus, the best scenario is to find a way to match an automated optical fibre placement system with the current industrial composite production technologies. Many different composite manufactures have indeed investigated the potential of an automated optical fibre placement structure that implement the control over embedding position, alignment, depth and pre-strain parameters. However, we note that one of the main issues related to employ an automated FOS placement embedding system is how to ensure its reliability and repeatability. Moreover, during the automated fabrication process, the integrity of delicate parts of FOS, such as the written grating area of FBG, and the control over the alignment of specific optical fibres such as MOF, are not fully guaranteed. Therefore, combining eventually a microcontrolled tomography based on X-ray with the automated-OFS system could be a relevant way for the integrated FOS quality control and an actual alternative for a convenient alignment of FOS for advanced and smart sensing operations.

References

[1]   W. Ecke, S. Grimm, I. Latka, A. Reutlinger and R. Willsch, *Proceedings of SPIE*, 2001, **4328**, 160.

[2]   I. Mckenzie and N. Karafolas, *Proceedings of SPIE*, 2005, **5855**, 262.

[3]   I.C. Song, S.K. Lee, S.H. Jeong and B.H. Lee, *Applied Optics*, 2004, **43**, 6, 1337.

[4]   C. Sonnenfeld, S. Sulejmani, T. Geernaert, *et al.*, *Sensors*, 2011, **11**, 2566.

[5]   J.A. Guemes and J.M. Menendez, *Composites Science and Technology*, 2002, **62**, 7–8, 959.

[6]   G. Luyckx, E. Voet, T. Geernaert, *et al.*, *IEEE Photonics Technology Letters*, 2009, **21**, 18, 1290.

[7]   H.K. Kang, J.W. Park, C.Y. Ryu, C.S. Hong and C.G. Kim, *Smart Materials and Structures*, 2000, **9**, 149.

[8]   A.K. Green, M. Zaidman, E. Shafir, M. Tur and S. Gali, *Smart Materials and Structures*, 2000, **9**, 316.

[9]   D. Kinet, P. Mégret, K.W. Goossen, L. Qiu, D. Heider and C. Caucheteur, *Sensors*, 2014, **14**, 7394.

[10]   L. Qiu, K.W. Goossen, D. Heider and E.D. Wetzel, *Optical Engineering*, 2010, **49**, 5, 054402.

[11]   S. Minakuchi, T. Umehara, K. Takagaki, Y. Ito and N. Takeda, *Composites Part A: Applied Science and Manufacturing*, 2013, **48**, 153.

[12]   C. Chojetzki, M. Rothhardt, J. Ommer, S. Unger, K. Schuster and H.R. Mueller, *Optical Engineering*, 2005, **44**, 060503.

[13]   N. Razali, M.T.H. Sultan, F. Mustapha, N. Yidris and M.R. Ishak, *The International Journal of Engineering and Science*, 2014, **3**, 7, 8.

[14]   H. Van, G.L. Bram, B. Erwin, *et al.*, *Sensors*, 2012, **12**, 12052.

[15]   J. Missinne, G. van Steenberge, B. van Hoe, *et al.*, *Proceedings of SPIE*, 2009, **7221**, 722105.

Index